Truthful Report on the Last Chances to Save Capitalism in Italy

By Censor
(Gianfranco Sanguinetti)

Translated from the French
by Bill Brown

Colossal Books
Brooklyn, NY

Published January 2014.

Colossal Books
POB 140041
Brooklyn, NY 11214

Translation copyright 2014 by William J. Brown.
All rights reserved.

Cover design: George Matthaei.

ISBN: 978-0-615-94827-0
Printed and bound in the USA.

Contents

Translator's Introduction	i
Dedication	2
Preface by the Author	4
I. Why Capitalism Must Be Democratic	10
II. How Capitalism Was Badly Managed in Italy	23
III. In Which the Social War Begins Again	32
IV. It Is Never Good to Merely Defend Oneself	45
V. What the World Crisis Is	59
VI. What the Communists Really Are	73
VII. Exhortation to Rescue Capitalism	81
Proofs of the Nonexistence of Censor by His Creator	100
Press Clippings Concerning Censor	113
Index	116

Translator's Introduction

"There is scarce truth enough alive to make societies secure; but security enough to make fellowships accursed. Much upon this riddle runs the wisdom of the world. This news is old enough, yet it is every day's news." William Shakespeare, *Measure for Measure*, Act III Scene II, lines 216-220.

The origins of Censor's *Truthful Report on the Last Chances to Save Capitalism in Italy* lie in the exchanges that Gianfranco Sanguinetti had with Guy Debord at the end of 1971, when these two men were drafting the documents that would eventually be published in *La Véritable Scission dans L'Internationale* (Paris: Editions Champ Libre, April 1972). Concerned with documenting the post-1968 history of the Situationist International ("SI"), which Debord had co-founded in 1957 and Sanguinetti had joined in 1969, and with continuing situationist subversion after the impending dissolution of the group, they struck upon the idea of publishing an essay titled "The Class Struggles in Italy."[1]

In the words of Debord's letter to Sanguinetti dated 3 January 1973,[2] such a text – now envisioned as a short book – would need to have the "trenchant" and "assured" tone of Machiavelli's *The Prince*. Under the heading "Notes on the book in progress," Debord made the following suggestions.

> 1. Italy before the crisis. The Italian *miracle* took place in a relatively backwards country, but within the platoon of the industrially advanced countries, and in the country that had *the strongest Stalinist party* in the West. Among the causes of the remarkable expansion of the Italian economy – linked to the global process – was the fact that Italy had a proletariat that was *deeply involved* (compared to the *Spanish* conditions of the same

---

[1] Cf. Debord's letter to Sanguinetti dated 13 December 1971: "On your side, do not forget that the proletariat, like the publisher, awaits your 'class struggles in Italy.'" *Guy Debord Correspondance, Volume 4, Janvier 1969 – Décembre 1972* (Librairie Arthème Fayard, 2004), p. 452. The phrase itself alludes to Karl Marx, *The Class Struggles in France, 1848-1850*.

[2] Published in *Guy Debord Correspondance, Volume 5, Janvier 1973 – Décembre 1978* (Librairie Arthème Fayard, 2005), pp. 14-17.

Translator's Introduction

period).

2. The origins of the crisis. The university and high school students of 1967, stimulated by the agitations of the rest of the world (the USA, Strasbourg) and clashing with much more archaic conditions (the *Zanzara* affair). Role of the influence of the Situationist International then and, later, when *the occupations movement* developed by agitating for factory committees in the north.

3. The Battipaglia process to the *hot summer* of 1969; then the hot autumn (here, the role of the Italian journal, the Venice Conference, your poster from November, etc.).

4. The bomb. What it *was*; what ends it served. Unfortunate story of Pinelli-Valpreda. The SI at that moment (*Reichstag*).

5. The ownership classes in their conscious and coordinated struggle against the proletarian revolution: the bourgeoisie and its disinherited younger sister – the Italian branch of the bureaucracy. Their negotiations for an "honest" division, among associates, of possession of Italian capitalism (*via* the State). How bureaucratic politics is difficult* and how its very success left it the poor share [of the spoils].

*One can only content the workers by buying them off; and if they are actually content, one no longer needs to pay them. But if one goes too far in contenting the workers, one risks completely losing control over them. And so it is necessary, not only to pay them, but to also accord a number of advantages to them (but according them *too much* radicalizes them, etc.)

6. The ripening of the crisis after the recoil due to the bomb. Reggio. The situation today (Agnelli and his anticipations, the attached article from *Le Monde*, etc.). The point reached by the F.A.I. in its worried hatred of the SI, a thousand other symptoms, without forgetting the revolts in the prisons, which have since spread to America and France.

7. What the proletariat wants; and how it can obtain it.

The central event in this chronology is "the bomb," that is to say, the bomb that exploded at the Piazza Fontana in Milan on 12 December 1969. Though the authorities defined the attack as an instance of terrorism and blamed it on anarchists (Giuseppe Pinelli and Pietro Valpreda were arrested shortly thereafter), the situationists believed that the attack was actually a "false flag"

Translator's Introduction

operation that had been planned and perpetrated by the secret services of the Italian State. On 19 December 1969, the Italian section of the SI published *Il Reichstag Brucia?*, which, as its title indicates ("Is the Reichstag Burning?"), likened the event to the Nazis setting fire to the Reichstag building on 27 February 1933 and blaming the attack on the Communists. Though they were virtually alone in making this claim, the Italian situationists were right: the bombing at the Piazza Fontana was the beginning of what later came to be called "the strategy of tension."

Sanguinetti took Debord's suggestions very seriously; one might even say that he took them literally. Not only does the *Truthful Report* contain seven chapters, but the contents of those seven chapters match up almost exactly with the contents of the seven sections in Debord's outline, as well.

But there is a crucial difference between the two. In Debord's outline, the person writing the history and analysis of Italian capitalism – the person attempting to show why the Italian State had recourse to a false flag operation against its own people – would be a real person (even if, like the authors of "Il Reichstag Brucia?" he didn't use his real name) and he would speak in his own voice and from the perspective of the situationist movement (the perspective of proletarian revolution). But the author of the *Truthful Report* is imaginary, a character who speaks words that have been placed in his mouth by a situationist. Furthermore, this character – he calls himself "Censor," a pseudonym that evokes the officer in ancient Rome who was tasked with supervising public morality and governmental finances[3] – is not an anti-capitalist revolutionary and he doesn't seek to destroy capitalism in Italy. He claims that he is a conservative member of the ruling class and that he wishes to save it.

Thus, unlike Debord's outline, which anticipates a "simple" subversion, Sanguinetti's *Truthful Report* is doubly subversive: in addition to using the truth to attack authority (not just "the authorities," but authority as such), it uses a usurped authority to do so. Significantly, though the idea to create Censor and use *him* to write the text of the *Truthful Report* was Sanguinetti's idea, he wasn't the first situationist to undertake a double subversion. In point of fact, it was the use of such techniques – the *détournement* (diversion) of

---

[3] In a letter to Mustapha Khayati dated 10 December 2012 and included in *On Terrorism and the State* (Colossal Books, 2014), p. 111, Sanguinetti says the name "Censor" was intended to echo "*Bancor*, [...] the supranational currency invented by Keynes" and that "it was also the penname of Guido Carli, who was the head of the Bank of Italy at the time."

Translator's Introduction

other people's ideas, the provocation of using them for subversive purposes, and the scandal caused by eventually revealing what one has done and why one has done it – that made the Situationist International such a powerful and effective organization, despite its small size, limited means, and short duration.

For example: in October 1966, the SI teamed up with a small group of radical students at the University of Strasbourg to publish *On the Poverty of Student Life*, which was a virulent attack on both French capitalism and the ineffectiveness of student protest movements. Written by a situationist (Mustapha Khayati)[4] and published under the auspices and with the funds of the official student union, *On the Poverty of Student Life* not only caused a major scandal in Strasbourg, but, thanks to the fact that it was widely distributed outside of Strasbourg and translated into several other languages, it also contributed to the growth of the revolutionary movement in France, Europe and the United States.

But Sanguinetti was certainly the first ex-situationist to use these situationist tactics after the dissolution of the SI. Thus he extended the subversion to a third level: not only did he use a usurped authority to attack authority as such, he also did so as an autonomous person, without the "authority" of the SI to back him up.

In the planning and execution of what came to be called "Operation Censor," Sanguinetti received help and encouragement from Ariberto Mignoli, who was his lawyer and friend, as well as from Guy Debord. According to "The Doge: A Recollection," which Sanguinetti wrote and published in 2012,[5] the character of Censor was based upon Mignoli (aka "the Doge"), and that, when Mignoli read the manuscript of the *Truthful Report*, he recognized himself in it. But though Mignoli was not an anti-capitalist revolutionary, he wasn't a supporter of Italy's ruling class, either. In Sanguinetti's words, "he scorned it as much as he knew it up-close."

Sanguinetti worked on the manuscript of the *Truthful Report* all through 1973 and 1974. A good deal of it must have been finished during those years because, in a letter dated 15 October 1974,[6] Debord told his friend that "the

---

[4] Guy Debord's contributions to the success of this scandal cannot be overlooked. Cf. his letters to Khayati dated 9 September 1966, 29 September 1966, 13 October 1966, and 19 October 1966, all of which are included in *Guy Debord Correspondance, Volume 3, Janvier 1965 – Décembre 1968* (Librairie Arthème Fayard, 2003), pp. 161-162, 164-165, and 165-168.
[5] Only available on-line. http://www.notbored.org/The-Doge.pdf.

Translator's Introduction

beginning of your pamphlet seems *magnificent* to me," that "all goes for the best" where "the tone, the dedication, [and] the pseudonym" were concerned, and that, "the Italian situation being what it is, I believe – fuck! – that this text could produce an effect much greater than the *Poverty* did in 1966." But Debord also warned Sanguinetti "it is necessary to finish work immediately," because there's always "the chance that the text might be rendered obsolete by new events."

Doing his best, Sanguinetti was still working on the manuscript in early 1975. "I prepared for [the publication of the *Truthful Report*] amidst a thousand dangers and unexpected events," he writes in "The Doge."

> In March of that year, I was imprisoned in Florence and charged by the principal Italian anti-terrorist prosecutor, Pier Luigi Vigna, on the very day that I was transporting the Censor manuscript to the printer in Milan. I was intercepted because the police had to know that I was preparing something and because Mignoli's phone was tapped because of the bankruptcy of a bank for which he was momentarily the attorney (at the time, I had no telephone as a precaution against taps). To arrest me, the police planted and 'found' bullets from a machinegun in the car in which I was traveling. The manuscript was saved because it had been placed in the violin case of my companion, Katherine Scott, who – along with my friend Mario Masanzanica – were also arrested. The manuscript thus had the singular luck of entering and leaving, unperceived, the women's prison at Santa Verdiana in Florence. The Doge furnished me with the best criminal-defense attorney in Florence, Terenzio Ducci, who, despite all expectations, got me out of prison in eight days.

The Italian political police were surveilling and harassing Sanguinetti for a number of reasons. First and foremost, despite the explosion of the bomb at the Piazza Fontana and other similar acts that were designed to intimidate them or portray them "terrorists," the revolutionary parts of the Italian working class had continued to go out on strike, to sabotage their places of work, and to receive support from other parts of Italian society. Second, despite the fact that the SI had dissolved in 1972, situationist ideas were an

---

[6] *Guy Debord Correspondance, Volume 5, Janvier 1973 – Décembre 1978* (Librairie Arthème Fayard, 2005), pp. 212-213.

essential part of the revolutionary movement. Finally, Sanguinetti himself was seen as one of the most dangerous situationists. He had been summarily expelled from France on 21 July 1971 for his membership in the SI; he had worked on the film version of Debord's book *La Société du Spectacle*, released in 1973; and, as he says, he was obviously preparing to do *something*, though the authorities didn't know what it was.

In June 1975, Sanguinetti finally finished working on his book. The following month, a Milanese printer by the name of Dario Memo set to work using the monotype process and special, high-quality paper to produce a luxury edition of Censor's *Rapporto veridico sulle ultima opportunita di salvare il capitalismo in Italia*. Only 520 individually numbered copies were made. Ostensibly published by Bergio Scotti-Camuzzi, who was in fact not a publisher, but Mignoli's cousin, the book was sent by mail in August 1975 to 520 Italian politicians, industrialists, union leaders and journalists, whose names and addresses had been furnished to Sanguinetti by Mignoli.

"We laughed heartily when we received [via Scotti-Camuzzi] the letters of thanks from government ministers and high-level civil servants, that is to say, all those who believed that Censor was real and sincere: Giulio Andreotti, Aldo Moro, Guido Carli (the governor of the Bank of Italy), Giorgio Amendola, Pietro Nenni, the Prefect of Milan, the High Council of the Magistracy, etc.," Sanguinetti recalls in "The Doge." The laughter intensified in October 1975, when the publisher Ugo Mursia brought out an inexpensive and widely distributed edition of the *Truthful Report*. In fact, this edition was so popular and so well reviewed in the Italian press that Mursia reprinted it twice over the course of the following two months.

We might well wonder how it was that Censor's book was not immediately recognized as a fake. After all, it *was* a fake, and, if one knew how to read between the lines (or even some of the lines themselves), it was an *obvious* one. Furthermore, fakes and their exposure had been in the news for several years before then. For example, in 1968, the Hungarian painter Elmyr de Hory, who had been forging dozens of paintings and selling them off to some of the most prestigious galleries and museums in the world for more than twenty years, was finally unmasked and imprisoned. In 1969, de Hory told his story to the American novelist Clifford Irving, who not only published *Fake! The Story of Elmyr de Hory, the Greatest Art Forger of Our Time*, but also, two years later, went on to perpetrate a fake of his own, the infamous *Autobiography of Howard Hughes*, for which he was imprisoned in 1972. To complete this cycle, in March 1975 the American film director Orson Welles released a feature-length film, *F. for Fake*, which documented the rise and fall of both de Hory and Irving.

It is obvious that fakers succeed by fooling the experts. Their fakes are "so good" that "even the experts" can't tell the difference. But Welles' contention was that this commonplace observation, i.e., fakers succeed *despite* the experts, has things backwards. In point of fact, fakers succeed *because* of the experts. It is because of the authority of experts – which is based upon the inability of everyone else to make educated decisions on their own and their consequent willingness to rely unquestioningly on the experts' judgments (which, in truth, are only their opinions) – that fakes are not only possible, but also highly lucrative when they are successfully perpetrated. Fool the expert, and you've managed to fool everyone else, in one fell swoop. In Welles' words,

> What's new? Experts are the new oracles. They speak to us with the absolute authority of the computer. And we bow down before them. They're God's own gift to the faker [...] "It's pretty but is it art?" How is it valued? The value depends on opinion. Opinion depends on the experts. A faker like Elmyr makes fools of the experts. So who're the experts? Who's the faker?[7]

Experts are merely authenticators, not creators in their own right. So as not to be exposed as the fakers that they truly are, they must be experts in two fields: their own particular area of expertise, whatever that may be (modern art or fine wines or political analysis); and the ability to dissimulate instances in which they have been fooled.

Not surprisingly, it is rare that the existence of fakes, once they have been discovered, is publicized. If a fake has been exposed after it has been widely accepted as the real thing, then one is entitled to wonder: how many other, similar things are also fake? But the people who have made purchases or decisions based upon the mistaken opinions of the experts do not want to find out the answer to this question. They prefer not to know, because, if the full extent of fakery were revealed, the entire market for a particular product might collapse and they would be wiped out.

Thus we have our answer. In Italy in 1975, the *Truthful Report* was not denounced as a fake because its author seemed to be an expert, someone who was well acquainted with many State secrets. He might have been bluffing, but what if he wasn't? If he said that the bombing of the Piazza Fontana had been

---

[7] Orson Welles, *F. for Fake* (1975). My transcription.

Translator's Introduction

perpetrated by the Italian secret services, and that didn't jibe with what you thought you knew about it, then maybe you weren't really privy to the truth. And yet the very fact that such an apparently knowledgeable person had taken you into his confidence seemed to be an indication that you were in fact a real expert. At least Censor recognized you as one. Any suggestion that his book was a fake opened you to the accusation that you were in fact not a real expert, but a fake one. And so, in the interests of maintaining your own status as an insider, you kept your doubts (if you had any) to yourself. And then, after the book was published commercially, no one among its secondary audience – ordinary men and women – was willing to proclaim that the thing was a fake because that would have contradicted the unanimous judgment of the "real" experts. And if *they* didn't think that Censor's book was a fake, then why should you, i.e., someone who lacked expertise, think otherwise?

By the same token, these insights about expertise explain how and why fakes and hoaxes have continued to exist and be successful. One would think that, precisely because we now live in a world in which more people have access to more information about more subjects, fakes and hoaxes would be impossible to perpetrate. For example, if you search on-line for the phrase "lying in the age of the Internet," you will see that the unanimous opinion is that it is no longer possible to lie: the widespread availability of information makes "getting caught" inevitable. But the simple truth is that, precisely because of the spectacular increase in the quantity of information, which requires time to wade through and evaluate, the bad quality of a lot of that information, and the speed with which it piles up, the numbers of and reliance upon experts have steadily increased. As a result, the numbers of successful hoaxes and fakes have increased proportionately. The year 2013, for example, has been called "the year of the hoax."[8]

In January 1976, Sanguinetti published an essay titled *Prova dell'inesistenza di Censore, enunciate dal suo autore*, which revealed that Censor did not exist and that he himself had written the *Rapporto verdico*. Precisely because no one had doubted the existence of Censor or had questioned the veracity of the claims that Censor had made (particularly where the Piazza Fontana and the State's use of "artificial terrorism" to stop

---

[8] Cf. Doug Gross, "2013: The Web's year of the hoax," published by CNN on December 18, 2013: "News alert: some things you read on the Internet are not true. As obvious as that may seem, and as savvy as you'd think we'd be a decade after deposed Nigerian princes began e-mailing us with the promise of vast riches, 2013 has turned out to be the Year of the On-line Hoax."

proletarian subversion were concerned), a major scandal ensued. In an attempt to make that scandal international in nature, Editions Champ Libre brought out a volume that included *Véridique rapport sur les dernières chances de sauver le capitalisme en Italie*, which was Guy Debord's translation of the text into French, plus Debord's translation of the *Prova dell'inesistenza di Censore* and selections from the book's extremely positive reviews in the Italian press.

In February 1976, calumnied by the Italian newspapers that had been so easily and completely duped by "Operation Censor" and hounded by the political police, who now found a new reason to harass him, Sanguinetti fled Italy. Even though he had been deported from France in 1971, he attempted to re-enter that country, where he had friends and supporters. Furthermore, the man who had been responsible for his expulsion, Minister of the Interior Raymond Marcellin, was no longer in office. But Sanguinetti wasn't allowed in, not even temporarily.

To bring this news to the attention of the readers of France's newspapers, which had not seen fit to publish a single word about it, Debord wrote a bitter and sarcastic statement on behalf of Editions Champ Libre. Published as an advertisement in the 24 February 1976 issue of *Le Monde*,[9] it focused on the failure of France's "post-modern" academics and the rest of the intelligentsia to register the existence of, not to mention denounce, the government's decision to refuse Sanguinetti entry.

> We do not have the presumptuousness to insinuate that the critique of capitalism could at all concern our contemporaries, their work, their ways of making a living, their ideas or their pleasures. We do not ignore the facts that, even as a subject for scholarly discussion limited to a small number of experts, the very justness of the concept of that critique has been controversial and that capitalism, as a hypothesis, is no longer of contemporary interest, because the Thought of [the Université de] Vincennes – at which the best-recycled professors have decided upon the dissolution of history and the prohibition of the criteria of truthfulness in discourse, which is something that is very rich in consequences for them – recently leapt beyond it.
>
> Furthermore, we are not assured that, somewhere, there

---

[9] Reprinted in Editions Champ Libre, *Correspondance*, Volume I (Paris, 1978), back cover.

Translator's Introduction

really exists a geographical (and an economically quite weak) entity called Italy. And, where Italy's economy is concerned, the eminent leaders of the Common Market – even if the principle of the free circulation of commodities is as much their affair as the free circulation of people – have other reasons to doubt its existence.

The actual existence of Gianfranco Sanguinetti himself – either as the author of a Western *samizdat* or as the target of some liberal-advanced Gulag – is highly questionable. If we, on the unique basis of the magnitude of a public rumor (which also remains outside of our borders), allow ourselves to positively affirm the reality of his existence, his writings and the diverse and harmless police persecutions that have followed from them, one could retort that no one here in France has ever heard of him, and we [as his publisher] feel all the weight of such an objection.

We will also frankly state that we know a number of estimable people who, working for the newspapers or the distributors of books, do not hide the fact that they have been led to conclude that Editions Champ Libre also does not exist, and, for our part, we do not pretend to have the boldness to settle such an obscure question and thus go against the honest convictions of so many competent people by basing ourselves only upon our contingent desires and limited personal interests.

Given all this, we nevertheless will not allow ourselves to leave open the question of knowing if the world in which we live – the world of which you read all the most up-to-date news every day – truly exists. We are in a position to be assured that, for the moment, it still does.

But Debord's words fell on deaf ears. Sanguinetti was left to fend for himself. He eventually returned to Italy, where, undeterred, he went on to write several new texts, including *On Terrorism and the State*, which was published in 1979.

* * *

Though it was published almost 40 years ago, the *Truthful Report on the Last Chances to Save Capitalism in Italy* is certainly worthy of being read and studied closely today. Historians of the situationist movement in general and

Translator's Introduction

the development of the thought of Guy Debord in particular will find it especially valuable. Though few have remarked this fact, the *Truthful Report* (and Sanguinetti's subsequent book, *On Terrorism and the State*) had a powerful influence on Debord's last major theoretical work, *Comments on the Society of the Spectacle*, which was published in 1988. Attentive readers will note a strong similarity between the five major historical developments presented in Chapter I of the former and the five principal features of "the society modernized to the stage of the integrated spectacle" presented in Chapter V of the latter.[10] For Debord, what had taken place in Italy in the 1970s was the harbinger of what was taking place in the entire world during the late 1980s because "the predominant place that Russia and Germany held in the formation of the concentrated spectacle, and that the United States held in that of the diffuse spectacle, seems to belong to France and Italy in the putting into place of the integrated spectacle, through the play of a series of shared historical factors: the important role of the Stalinist party and unions in political and intellectual life, a weak democratic tradition, the long monopoly on power by a single party of government, and the necessity to put an end to unexpected revolutionary contestation."[11] For Debord, it had been in France and Italy that the terrorism – "artificial terrorism," which isn't violence perpetrated against the State by extremists, but violence launched by the State against itself or the population that it rules – that came to dominate the entire world was first practiced.

The *Truthful Report* was also important to one of Sanguinetti's friends, Pier Franco Ghisleni, who used the tactic of usurped authority to generate *Lettere agli eretici: Epistolario con i dirigenti della nuova sinistra italiana* ("Letters to the Heretics: Correspondence with the Leaders of the New Italian Left"). Not only was this satirical work attributed to Enrico Berlinguer, the head of the Italian Communist Party, but it also presented itself as if it had been printed by the publishing house founded and run by Giulio Einaudi. Though the *Lettere agli eretici* did not create the immense scandal that was caused by the *Truthful Report*, it did create a minor sensation.

What about writers who were not members of the SI or one of Sanguinetti's friends? As Sanguinetti himself has said, "The Situationist International is historically confirmed as a true avant-garde only to the extent that its practices and theories have been applied, taken up, developed, détourned, publicized, etc., by other groups and individuals in other forms, situations,

---

[10] Guy Debord, *Commentaires sur la société du spectacle*, 1992, p. 25.
[11] *Ibid.*, p. 22.

Translator's Introduction

conditions, etc."[12] Though they may or may not have modeled their respective actions on the *Truthful Report* in particular, all of the following contemporary individuals or groups have certainly been explicitly inspired by the SI's use of détournement, provocation and scandal: the American activist group the Yes Men, whose members create fake websites and pretend to be corporate spokesmen; the American pro-privacy group the Surveillance Camera Players, which compiles and releases maps of the locations of publicly installed surveillance cameras; the English artist Banksy, whose graffiti art is a form of vandalism; the Russian punk band Pussy Riot, which plays scandalous songs in provocative settings; and the Czech art group Ztohoven, which hacks into State TV broadcasts and official government ceremonies.

But what about the contents of the *Truthful Report*? Even though it recorded the history of Italy between 1943 and 1975, and even though it made history in 1975 and 1976, this book is virtually never mentioned in discussions of State-sponsored terrorism or "false flag operations," even when anti-capitalist revolutionaries hold those discussions. The same goes for Sanguinetti's *On Terrorism and the State*. Neither book is mentioned in any of the many studies that have been published on these subjects,[13] nor are they

---

[12] Email message sent to me on 14 August 2012.
[13] Kenneth R. Langford's *An Analysis of Left and Right Wing Terrorism in Italy* (Defense Intelligence College, 1985); Leonard Weinberg and William Lee Eubank's *The Rise and Fall of Italian Terrorism* (Westview Press, 1987); Richard Drake's *The Revolutionary Mystique and Terrorism in Contemporary Italy* (Indiana University Press, 1989); Robert C. Meade's *Red Brigades: The Story of Italian Terrorism* (Macmillan, 1990); Raimondo Catanzaro's *The Red Brigades and Left-wing Terrorism in Italy* (Pinter, 1991); Marco Rimanelli's *Waning Terror: Red Brigades and Neo-Nazi Terrorism in Italy* (World Jurist Association, 1991); Jeffrey McKenzie Bale's *The "Black" Terrorist International: Neo-fascist Paramilitary Networks and the "Strategy of Tension" in Italy, 1968-1974* (University of California, Berkeley, 1994); Paul Ginsborg's *A History of Contemporary Italy: Society and Politics, 1943-1988* (Palgrave Macmillan, 2003); Ganser Daniele's *NATO's Secret Armies: Operation GLADIO and Terrorism in Western Europe* (Routledge, 2004); Silje Dalsbotten Aass's *State Responses to Terrorism in Italy: The Period 1969-1984* (S.D. Aass, 2005); Graeme Allen Stout's *Arrested Images: Discourses of Terrorism in Italy and Germany* (University of Minnesota Press, 2006); Anna Cento Bull's *Italian Neo-Fascism: The Strategy of Tension and the Politics of Non-Reconciliation* (Berghahn Books, 2007); Pier Paolo Antonello's *Imagining*

Translator's Introduction

cited in any of the Wikipedia entries for "Operation Gladio," "Gladio in Italy," "the strategy of tension," "the years of lead," "false flag" or "state terrorism," or in any of the archives maintained by libcom.org, a libertarian Marxist website.

Perhaps the reason for the spectacular absence of references to and discussions of Sanguinetti's books is that copies of them are hard to come by. Small presses published both the *Truthful Report* and *On Terrorism and the State* and, as a general rule, book reviewers and the book-buying public ignore the publications of small presses. But it seems that something else is at work here and, in fact, has been at work for many years.

In January 1980, in his "Preface" to the French edition of *On Terrorism and the State*,[14] Sanguinetti himself noted "the quasi-complete silence that has surrounded a book [*On Terrorism and the State*] that deals with a subject that is spoken about every day, but always in the same mendacious way, on the front pages of all the Italian newspapers as well as on the State-sponsored radio and television stations" and that the existence of his book has been "kept secret by the very people who are believed to have the obligation to speak about terrorism."

The reason for this silence is, I believe, easy to imagine. Sanguinetti didn't simply assert what many people refused to believe at the time, namely, that the Italian State had bombed, wounded and even killed some of its constituents. He also denounced those who refused to believe that such a thing could ever happen. And these people, and all those for whom they spoke, never forgave him, even though – or precisely because – history has proved that he was right. Such is the price for proving that the experts have lied: they lie about you; they deny that you even exist.

* * *

The mistake that the Italian secret services made – the mistake that made "Operation Censor" possible – was that they turned the tactic of artificial terrorism into a strategy. Instead of using it sparingly and only when absolutely necessary, they began to use it again and again. As a result, they risked the long-term loss of everything that they had gained in the short term

---

Terrorism: The Rhetoric and Representation of Political Violence in Italy 1969-2009 (MHRA, 2009); and Richard Cottrell's *Gladio, NATO's Dagger at the Heart of Europe: The Pentagon-Nazi-Mafia Terror Axis* (Progressive Press, 2012).

[14] *Du Terrorisme et de l'etat*, translated from the Italian by Jean-François Martos (Le fin mot de l'Histoire, Paris, 1980), pp. 5-6.

Translator's Introduction

at the Piazza Fontana.

Perhaps this is why there have been no instances of artificial terrorism in the United States since September 11, 2001: the State knows the risks that it runs if it over-uses it. And from the perspective of the State, the false flag operations undertaken on September 11 were completely successful.

But what if the situation in China[15] gets out of control? What if the

---

[15] Cf. Eli Friedman, "China in Revolt," *Jacobin Magazine*, Issue 7-8, August 2012, from which I quote at length because of the very strong similarities between the situations in Italy in the early 1970s and in China today.

"Today, the Chinese working class is fighting. More than thirty years into the Communist Party's project of market reform, China is undeniably the epicenter of global labor unrest. While there are no official statistics, it is certain that thousands, if not tens of thousands, of strikes take place each year. All of them are wildcat strikes – there is no such thing as a legal strike in China. So on a typical day anywhere from half a dozen to several dozen strikes are likely taking place. More importantly, *workers are winning*, with many strikers capturing large wage increases above and beyond any legal requirements. [...] Strikes [...] are never organized by the official Chinese unions, which are formally subordinate to the Communist Party and generally controlled by management at the enterprise level. Every strike in China is organized autonomously, and frequently in direct opposition to the official union, which encourages workers to pursue their grievances through legal channels instead. [...] When faced with recalcitrant management, workers sometimes escalate by heading to the streets. This tactic is directed at the government: by affecting public order, they immediately attract state attention. Workers sometimes march to local government offices or simply block a road. Such tactics are risky, as the government may support strikers, but just as frequently will resort to force. Even if a compromise is struck, public demonstrations will often result in organizers being detained, beaten, and imprisoned. Even more risky, and yet still common, is for workers to engage in sabotage and property destruction, riot, murder their bosses, and physically confront the police. Such tactics appear to be more prevalent in response to mass layoffs or bankruptcies. A number of particularly intense confrontations took place in late 2008 and early 2009 in response to mass layoffs in export processing due to the economic crisis in the West. As will be explained, workers may now be developing an antagonistic consciousness vis-à-vis the police. [...] A turning point came in the summer of 2010, marked by a momentous strike wave that began at a Honda transmission plant in Nanhai.

Translator's Introduction

workers of the United States and Europe – inspired by their counterparts in India and China – become rebellious again? What if critical mass is reached in the clamor to reign in or even *cease and roll back* the systematic surveillance of the world's populations by the United States' intelligence agencies, which, of course, has been done in the name of "fighting terrorism"?

Then Western capitalism might find that one or two more instances of artificial terrorism are "necessary" for its survival. If it does, then a new "Operation Censor" will become possible. I do not relish such a possibility; I simply hope that the revolutionary movement of the future will have one of the weapons that it will need when history starts repeating itself.

* * *

A few notes about the text. Since I do not speak Italian, I have used Guy Debord's *Véridique rapport sur les dernières chances de sauver le capitalisme en Italie* as the basis for this translation into English. The Italian original included words and phrases from a number of other languages (mostly Latin, French and English). Debord was careful to preserve this multi-lingual richness as he translated the work as a whole from Italian into French, and I, translating from French into English, have tried to be careful, too. When Censor has quoted from something in English, I have sought out and used the original wording. When Censor has quoted from something in Latin, I have consulted and relied upon the already-established rendering of it into English. All of the footnotes are by me, except where noted. The typographical imprint on page 1 reproduces the 17th century woodblock engraving that Ariberto Mignoli chose for publication in the colophon of the first numbered edition of the *Truthful Report*. Finally, this edition of Censor's pamphlet is the first to include an index of the important names, events and places mentioned in the text.

Bill Brown

Since then, there has been a change in the character of worker resistance, a development noted by many analysts. Most importantly, worker demands have become *offensive*. Workers have been asking for wage increases above and beyond those to which they are legally entitled, and in many strikes they have begun to demand that they elect their own union representatives. [...] In just a few years, worker resistance has gone from defensive to offensive. Seemingly small incidents have set off mass uprisings, indicative of generalized anger."

Truthful Report on the Last Chances to Save Capitalism in Italy

Gianfranco Sanguinetti

To the amicable memory of Raffaele Mattioli, who taught us to be lavish with the most precious of our goods: the truth.

Truthful Report on the Last Chances to Save Capitalism in Italy

Then it replied: 'A conscience that is clouded
By its own shame or by that of another,
Will certainly feel that your words are sharp.

But none the less, all lying set aside,
Make clear to everyone the whole vision;
And let them scratch wherever they may itch.

For if you words are objectionable
For the first taste, they will yield nourishment
Afterwards, once they have been digested.

This cry of yours will do as the wind does,
Strike hardest on the summits that are highest;
And that is no small argument of honor.'

Dante, *Paradiso*, XVII, 124-135.

Gianfranco Sanguinetti

Preface

The author of this *Report* is afflicted with a great disadvantage: it seems to him that nothing, or almost nothing, must be treated in a light tone. The 20th Century thinks the opposite, and it has its reasons for this. Our democracy, which demands the expression of personal opinions from an infinity of brave people who do not have the time to form a single one, forces everyone to speak with a thoughtlessness that we, in our turn, are obliged to excuse, given the necessities of the times.

Nevertheless, this first disadvantage does not shelter us from an opposite one: if we refuse to use a light tone, we also reject an academic or serious style for the good reason that we do not intend to demonstrate in 50 pages what can be said in five lines. We hope that this double premise will at least serve to excuse, if not justify, the *trenchant*[1] tone.

In these first few lines, we would like to thank a number of illustrious Italians, whom we would name if they were dead, but who at this moment are occupied with important tasks in our economy and politics, and thus will be grateful to us for our discretion, given the undeniably delicate character of the subjects treated herein. All that we can permit ourselves to do is offer to them these pages, which we have finally decided to publish under the rubric of this *Report*, although, we must confess, we secretly but unsuccessfully nourished the hope that someone other than us would undertake it. On the other hand, given the speed of the Italian crisis, and the urgency of adopting remedies, we have had to resolve ourselves to confiding our opinions in a published work, because, after their previous distribution in the form of confidential notes and private conversations, it hasn't seemed to us that they have encountered all of the desired audience, precisely "there where one can do what one pleases"[2] that is to say, at the summit of economic power.

It is fitting to say immediately that we do not intend to speak for *all* of the Italian bourgeoisie, which has been bastardized by its own illusions of "openness," but only a part of it, in which one can distinguish a veritable *elite*[3] of the powerful. It is to this elite that what follows is addressed, in an epoch in which the monopoly on the more or less critical discourse on contemporary

---

[1] French in original.
[2] There "where everything is possible." Dante, *Inferno*, V, 23-24, *The Divine Comedy*, translated by C. H. Sisson (Oxford University Press, 1993).
[3] French in original.

Truthful Report on the Last Chances to Save Capitalism in Italy

society seems to belong to those who are opposed to it in a more or less effective manner, while on our side of the barricade one discerns a pitiful silence and even an ever-more clumsy recourse to embarrassed justifications for it. As for us, at this moment in which we break this monopoly, we are quite far from wanting to seek the least appearance of "dialogue" with our real enemies. We speak to the heart of our own class so as to perpetuate its hegemony over this society.

Unlike those who critique society so as to revolutionize its bases, we will not make grand demagogic or pedagogic speeches; and rather than responding to our radical critics, we prefer to personally assume the *disgraced grace*,[4] that is to say, the displeasing honor of criticizing, even pitilessly, that which in our management of economic and political power must be effectively criticized with the sole goal of reinforcing efficiency and domination.

Thus we do not seek to prove that contemporary society is *desirable*, and even less to weigh the possibly modifiable aspects that compose it. With all the cold veracity that we have adopted for all the other affirmations contained in this *Report*, we say that *this society suits us because it exists* and we want to maintain it to maintain our power over it. To speak the truth in these days is an exacting and time-consuming task, and since we cannot hope to exclusively encounter impartial readers, we will content ourselves by being ourselves as we write, even at the price of making accusations against the politicians who, over the years, have defended our interests with more good will than success. We must cease to be hypocrites amongst ourselves, because we are in the process of becoming victims of this hypocrisy.

Today, from the point of view of the defense of our society, there only exists a single danger in the world, and it is that the workers succeed in *speaking to each other* about their conditions and aspirations *without any intermediaries*. All the other dangers are attached to, or even proceed directly from the precarious situation that places before us this primary problem, which in many respects is concealed and unacknowledged.

Once this true danger has been defined, it is a question of exorcising it, and not seeing false dangers in its place. Yet our politicians only seem preoccupied with saving their own reputations, and too often this comes too late. But, on the contrary, it is *saving our basis,* which is economic above all, with which they must occupy themselves. For example, we have noted the stupidity that currently dominates the debate, conducted under the heading of "the Communist question," among the principal political leaders, as if this

---

[4] Greek in original.

were a problem that was so embarrassing as to be "new," as if we ourselves – and several others, who are certainly no less qualified – had not already set the form, timing and conditions that will render the official entrance of the Italian Communist Party [ICP] into the sphere of power useful for both sides, and as if the Communist leaders had not, during the most recent meetings that we have held, already unofficially accepted even the most unfavorable aspects of the project that at this moment, with the prudence that is now necessary, they are attempting to get the rank and file of their party, which believes itself to be the most radical, to accept. This imaginary political debate, which does not even serve the majority parties by assuring them of the support of moderate voters – which is a superfluous concern, since the voters always vote as they are told to vote – , cannot mislead the intelligent conservatives, either in Italy or abroad, because we know that it is no longer a question, at the current moment, of seeing if we more or less need the ICP, given that no one can doubt the utility that this party has been to us during the last few, very difficult years, when it would have been so easy for its leaders to harm us and perhaps in an irreparable fashion, but instead a question of us being in a position to offer this party sufficient guarantees so that it will not run the risk – once it is openly allied with our management of power – of being involved in our possible ruin, for which the ICP would *ipso facto* find itself sharing the responsibility and the consequences by, at the same time, losing its own basis among the workers who, no longer having any illusions about the most minimal changes in their fate – a fate that is indeed hardly enviable – and no doubt estimating this to be a betrayal by their leadership, would react freely, beyond any control and against all control. That's the real question; that's the real danger.

We know quite well that the Communist parties have many times furnished proof of their aptitude at collaborating in the management of bourgeois society, but we must not rely on such a general certitude, as if it would confer upon our power a reserve of unlimited security, that is to say, a recourse that would be sufficient *in every case* no matter what "the day and hour" of the supreme danger would be, as if this recourse would not itself be a historical force among others, as if this force wouldn't be susceptible of *wearing out*, either through inaction or an action that was too maladroitly or too tardily engaged in. The height [of folly] for us would be finding ourselves, precisely ourselves, to be the last dupes of the Communist myth by betting on the *fantasy of its omnipotence*, which we ourselves have supported at the times in which it was advantageous for us to combat it. We must never forget that *the only effective power* is ours, and that it is nevertheless threatened.

Truthful Report on the Last Chances to Save Capitalism in Italy

Thus, it isn't sufficient to know that the Communist Party is ready to manage society for our profit; we must also have a place to offer it in a capitalist society *that still merits being managed.* Who doesn't understand that, if the State and civil society continue to deteriorate at such a dramatic speed under the pressure of truly irreconcilable enemies whom we – the Communists and us – have *in common,* the Communists, caught up with us in the same disaster, will find themselves as incapable of helping us as the Austrian-Hungarian Empire or the Kingdom of Jerusalem? If, at that moment, the Communists deplore the fact that they can no longer maintain the existing order, that will be a subjective event that will not offer us any consolation! And if the Communists, by once again taking up the weapons of counter-revolution, crush the attempt to set up a classless society in Italy, they would certainly merit the recognition of the property-owning classes in America and Russia, in Europe and in China, and they could be admitted more or less quickly into the UN as the masters of our country, but we – the real dominant class in Italy, the particular class that can even call itself the founders of the universal bourgeoisie of modern times and the *millennium* that it has effectively imposed on the entire world – will no longer be here. We will endlessly experience *how salty is the taste*[5] of the bread of exile in London or Madrid.

What we must save isn't only the capitalism that maintains the market economy and salaried employment, but, rather, capitalism *in the only historical form that suits us,* which, moreover, can quite easily be shown to be the effectively superior form of economic development. If we don't even know how to offer the Communists a *chance* to save this form of capitalism, they will confine themselves, as much as they can, to saving *another form of it,* the unfortunately rustic character of which one has seen in Russia for more than a half-century. The new class of property-owners that this inferior form produces, one knows well, leaves us no existence locally, just as it also suppresses – everywhere in which its crude dictatorship takes the place of the one that we don't fear to call ours – the totality of the superior values that give existence a meaning.

What we have said here are banalities, obvious facts. Those who do not accept them are sleepwalkers who haven't for a moment reflected on the fact we will lose all of our reasons for managing a world in which our objective advantages have been suppressed from the moment that it will no longer be possible for anyone to enjoy them. Capitalists must not forget that they are also human beings, and as such they cannot accept the uncontrolled

---

[5] Dante, *Paradiso,* XVII, 58.

degradation of *all* human beings and thus the personal conditions of life that they especially enjoy.

We would like to prevent an objection, nay, a reproach, that could be addressed to us, and that we judge to be absolutely unfounded when it comes to our *Report*: namely, that we herein reveal secrets that we have come to know over the last few years, which, when it comes to State secrets, have certainly not been few and far between, and that we divulge them without preoccupying ourselves with the possibly dangerous consequences they will cause in public opinion. Well! We can immediately reassure anyone who fears this: if one takes into account the double presupposition, which is too neglected in our country, that, on the one hand, he who always lies will never be believed, and, on the other, the truth is destined to forge its route with a force that can override the most powerful lies, whose destiny it is, on the contrary, to lose all of their strength when and to the extent that they are repeated, then we will see that the small number of naked truths that we have decided to reveal in this pamphlet can no longer be kept quiet without our running the risk that, in a short period, one or another of them will be put to seditious ends.

Moreover, our remarks will be quick, and we will never dwell on anything for too long, supposing that the readers to whom we are addressing ourselves through special means, and who are the very people with whom we have done business during these last few years, are sufficiently up-to-date concerning a good part of the delicate details, of which we will content ourselves with a quick review, that they will grasp the insinuations or allusions to facts or individuals, while all this will completely escape those who live at a distance from the centers of power in our society.

Instead of the celebrated phrase "I am prohibited from speaking and I cannot keep quiet,"[6] we prefer the honesty of "I will not say everything, but everything that I say will be true."[7]

*

Perhaps it might not be useless for me to specify, before concluding this preface, that we are not in the habit of writing books, not because we don't love reading them, but precisely because we love them more than this century seems to permit us. This is why, personally speaking, we are grateful for those

---

[6] Latin in original.
[7] Latin in original.

Truthful Report on the Last Chances to Save Capitalism in Italy

who today *do not write them* and we abhor the amateur or professional writers of our times, in which illiterate intellectuals unsuccessfully pursue the remission of their ignorance by publishing the proofs of it in a multitude of unreadable volumes, volumes that our culture industry undertakes to erect as a kind of barricade against true culture, which is currently out of fashion. If we ourselves have taken up the pen, this should rather be interpreted as our manner of payment of a *unique*[8] tax to the troubled Republic. And, if we have wanted to give to this *Report* the literary form of the pamphlet, which has been out of fashion for two centuries, this is only because it possesses the double advantage of being easy to read and quick to write. In it we address ourselves to men for whom the time to read is less than the necessity to act. And if we ourselves reject the method of reading quickly what appears to be important, without exhaustively treating each question that is raised, perhaps we might leave behind some monumental work of which the historians will one day make use to shed light on the years in question here, but in such case we would lack the time to confront and master (such is our intention) the crucial problems that we limit ourselves herein to sketching out, because we are not in the habit of believing that real difficulties can be resolved *through writing*. Thus, this pamphlet must be read as it was written: in one sitting, following the mood of the moment – a mood that, in this case, cannot be better than the gravity of the moment allows.

As for the fact that the author of this text has used a pseudonym: this was done to respect the tradition of the pamphleteer, illustrated by the Fronde under Mazarin[9] and by Junius in 18th Century England. Moreover, we are sure to be easily recognized by all those who have had the occasion to encounter us over the course of the last 30 years. Finally, for all the others, we prefer that it isn't our name that encourages the most rigorous reflection, but the seriousness of what we evoke.

Censor
June 1975

---

[8] Latin in original.
[9] France between 1649 and 1652.

Gianfranco Sanguinetti

Chapter I:
Why Capitalism Must Be Democratic and the Grandeur It Achieves By Being So

"You will soon be, thank Heaven, out of the hands of your rebellious subjects (...) Where they are concerned, my Cousin, I share all of your feelings, as you can see, and pray God that that He will keep you safe, but I cannot approve of your repugnance for the type of government that one calls representative and that I myself call recreational, there being nothing in the world that is so entertaining for a king, not to mention the not insignificant utility that it has for us (...) The representative form of government suits me marvelously (...) Money comes to us in abundance. Ask my nephew in Angoulême [in France]. Here we count by the thousands or, to tell the truth, we ourselves no longer count, because we have our own representatives, a *dense* majority of them, as one says here; expenses, but they are small (...) One hundred voices, I am sure, doesn't cost me in a year what Mme. de Cayla costs in a month (...) I truly thought as you did, before my trip to England; I had no love at all for representative government; but there I saw what it really is. If the Turk suspected as much, he wouldn't want anything else, and he would make his Divan a two-chambered body (...) You shouldn't be scared off by the words liberty, the general public, or representation. They work to our benefit, and their products are immense, the danger nonexistent, whatever one says...."

(These extracts, translated into Italian here for the first time, come from a secret letter that Louis XVIII sent to Ferdinand VII in August 1823. In Cadix, this letter fell into the hands of a secret agent from Canning, and its publication caused a controversy in England. – *The Morning Chronicle*,[10] October 1823.)[11]

---

[10] English in original.
[11] As Sanguinetti would later point out in his text "Proofs of the Nonexistence of Censor by His Author" (December 1975), "the letter attributed to Louis

Truthful Report on the Last Chances to Save Capitalism in Italy

What constitutes the most notable trait of our century isn't so much the fact that capitalism has been challenged in a reiterated and bloody manner by the workers of all industrialized countries and also in some countries where the economy is still predominantly agrarian (not at all unexpected phenomena, except to those who undervalued the warnings issued by the first failed revolutions of the prior century), nor the fact that serious economic and monetary crises have regularly shaken internal stability (serious inconveniences, but unavoidable in any complex economic system), nor even the fact that errors in the management of power have been quite numerous and very costly in every country (this fact is inseparably tied to any historical form of domination). It seems to us that what is notable in our century, quite the contrary, is that the capitalist system has managed to resist all that, and that, *despite all that,* today it still continues to exist everywhere, in manifestations that are different *and even appear to be contradictory,* as the only existing form of domination in the world, not only capable of surpassing its own crises, but even coming out of them reinforced to the point that it has managed to spread and impose its methods of production, exchange and commodity distribution upon the whole planet. Even in the Communist countries, the economic and technological systems of modern capitalism have long since become the declared preference of the dominant bureaucratic class.

For the first time in universal history, a determined system has imposed itself *everywhere,* annihilating all of the archaic forms of domination that were opposed to it, at the same time that it has successfully confronted the questions posed to it by new social forces, such as the class of industrial workers and salaried workers in general, who are necessary for the production and consumption of commodities, but who have an underlying disposition to combat in the name of their own "emancipation" the world for which they work and in which they live.

At the beginning of a *Report* dedicated to the critique of the current management of our system, it appears to us necessary, and just, to recognize its unquestionable historical success and its objective merits, which we risk seeing compromised in the near future because of current errors. It is fitting to know clearly *what to preserve* in what we must fight *here and now,*[12] and to be aware of what we have to lose at a moment when it is indispensible to

---

XVIII is in fact a celebrated literary fake by Paul-Louis Courier."
[12] Latin in original.

choose how to comport ourselves, and what weapons will help us, if we wish to emerge victorious from the very grave crisis that is the cause of our worries and the origin of this text.

According to Thomas Carlyle, the French Revolution had the demand for truth as its essential meaning. It was an historic proclamation of the fact that all lies, on which one had up until then based the harmonious organization of a social hierarchy, had to be rejected from then on. If these ideas are correct, we can determine that, for the last two centuries, we should have been able to avoid the greatest part of what harms us.

All of the historically dominant forms of society have been imposed on the masses, who quite simply must be *made to work,* either by force or by illusion. The greatest success of our modern civilization is that it has been able to place an incomparable *power of illusion* at the service of its leaders. Later in this pamphlet, we will see that this is also where the weakness of our power lies and threatens to become a serious crisis at any moment, because this illusion must *never* be shared by the ruling *elite*[13] that produces and makes use of it. Accumulative and rapid economic development (accumulative in the dimension of its rapidity), as well the positive technological upheaval that incessantly accompanies this development as its corollary, have caused in the totality of production and distribution an extreme concentration and a control that tends to become absolute. What has unfortunately challenged the current state of the world is the fact that this control possesses a strategy on the scale of its immense means. We will return to this point. But what is beyond doubt is the fact that economic development itself has demanded and brought about (in previously unimaginable proportions) the separation and passivity of the agents of production, that is to say, the very same ones who are identified by another branch of the social sciences as "consumers" and "citizens."

This situation has produced, as a natural product of our stage of historical development, *the social necessity for contemplation,* which Bergson, in his time (in the pages of *Creative Evolution*), called "a luxury." This contemplation is opportunely satisfied by the privileged part of our technology that is dedicated to the fixation and diffusion of images. The reason for this cannot escape anyone of good faith. The objective and measurable successes of our society are completely economic and technical. What this society produces is more things to watch. Some people have asked us, moved by perfectly irrelevant sentimentality: "Must we also love this society?" The question is asked in vain or, rather, if one admits that posing

---

[13] French in original.

Truthful Report on the Last Chances to Save Capitalism in Italy

such a question from any transcendent point of view means that real society would be a pure absurdity, we can only say that the question is effectively asked in vain in the sense that it has already fully found its response from the moment that one poses it in terms of real society, that is to say, in terms of social classes, by wondering, "*Who* must love this system of production?" Those who appropriate surplus-value necessarily love the existing form of production. As for the others, why should they love it? Production in itself appears to them as a simple necessity, and this is what it really is. As for the particular form this necessity assumes, those who hold capital don't find it any more defensible than any other form, and are only attached to it due to the specific advantages that they draw from it. If the excessive hypocrisy of the social thought of our epoch hadn't so mixed up and dirtied the playing cards that, cheating as always, it has ended up being unable to cheat intelligently, we would blush to recall such truisms. Our workers have in no way decided upon what they produce. And this is quite fortunate, because we might wonder what they would decide to produce, given what they are? It is quite sure, whatever the infinite variety of conceivable responses, that a single truth would be constant: they would assuredly not produce anything suitable for the society that we manage. And as these workers cannot be dazzled (no more than you or we ourselves) with happiness by the enlargement of the organizational chart of a multinational corporation or by the rate of growth in the sales of fighter planes to the Middle East, but find themselves deprived of any real compensation in the existence that is created for them, we must distribute to them some other compensation. This is what is accomplished by the massive diffusion of images that can be contemplated, though they no longer constitute the "luxury" spoken of by Bergeson, but a contemplative necessity, a *diversion*[14] like the Roman *circuses*[15] or Pascal's definition of the term.

Whatever the importance, and even the gravity, of the dangerous weaknesses of our power that we must criticize today, we must not lose sight of the fact that all this is subordinate to these brilliant successes. One only defends a social order that is alive. And if bourgeois society hadn't won this victory at the universal level, we wouldn't be here today to discuss its defense, because it would otherwise be as dead as Darius' Empire.

If we take a moment to remember (and that would be a healthy propaedeutic during the current struggles) that, for the last hundred years, we

---

[14] French in original.
[15] Latin in original.

have run the risk of having the world escape from of our grasp in a short period of time, we will ascertain the importance of the reprieve that we have obtained, which, in addition, has permitted us to undertake a profound transformation of all the conditions for this strategy – a transformation that we can define as follows: the construction of a new terrain of battle on which we await a disoriented adversary who must at first *recognize* it as such and then is constrained to advance while surrounded by the powerful defenses that we have wisely set up.

One can say that the 19th century, in the wake of the frightening revolutions of 1848, discovered political economy. Society divided into classes and private property had already been challenged: the critique of them seemed inexorably tied to the progress of knowledge, notably among the working classes. Thus, because the ruling class feared the education of the working classes and universal suffrage (and apparently quite legitimately so), it tied its defense to a position in the past, to an attitude of retreat, which continually became more pronounced. Modern industry required education, at least a summary one, and education, by spreading, necessarily worked in favor of universal suffrage. The bourgeoisie remembered that the progress of its leading lights had accompanied its own march to political power, and it feared that the same route would be followed by the proletarians. Fortunately, the proletarians also believed in this identification of their respective destinies; both classes thereby deceived themselves, because the two revolutionary projects were so different that they could not make use of the same leading lights, nor their diffusion and usage by analogous means. Thus, both the fears of one class and the hopes of the other were in vain.

Over the course of the century, the development and expansion of political and economic power changed the face of the world, much more than any past revolution had been able to do. What have been the characteristics and the permanent effects of this change? What did it destroy, and what did it create? It seems to us that the moment has come to define and set forth the distinctive traits of the new reality, because today we find ourselves at the precise point where we can best evaluate the results of a series of upheavals. Though we are far enough from their beginnings to be sheltered from the passions of those who began them, we are close enough to them to distinguish their essential elements. Soon it will be difficult to make an objective judgment of these events, because, by making their causes disappear, the great historical changes that succeed subsequently become less comprehensible due to the very fact of their success. Thus we will now consider the secrets of our victories in the old campaigns, not to seek some

Truthful Report on the Last Chances to Save Capitalism in Italy

hollow compensation in our pride in the successes of bye-gone days, but rather, at the heart of a new war that has suddenly been revived throughout the entire social field, to pull together and consciously use these secrets in other battles that we are called upon to fight anew. In the epic tale of the old social war, what were our decisive battles, our Salamines and our Marengos?[16]

To be brief, we will distinguish five of them.[17]

First, we have in a certain manner challenged Carlyle's remark by quantitatively and qualitatively realizing *the progression of the lie in politics* to a degree of power never before seen in history, with its content growing alongside the proliferating extension of its means. It developed with the "radical" bourgeoisie and its journalistic and parliamentarian practices, which followed the workers' movement organized as socialist political parties. The process begun by the parliamentary representation of the citizens has been quite naturally and considerably reinforced by the success of the unionized representation of the workers, since it is true that *all* representation plays our game. What one has customarily called *brainwashing*,[18] that is to say, the propaganda of false news diffused day after day by all the governments during World War I, has subsequently crossed a threshold beyond which, in normal times, one wouldn't have believed it possible to take literate citizens. Cardinal Carafa's remark, made at the time of the Inquisition, remains true: *People who want to be deceived will be deceived*.[19] Fascism was a pathological excess of the unlimited lie, but also a remedy in a time of crisis. But it is fitting to note that fascism completely failed due to its very nature, but by no means on the terrain of its means of propaganda, to the point that Hitler could theorize the fact that "the masses . . . will be more easily deceived by a big lie than by a small one." The advertising of the modern market then came to exploit the possibilities more rationally, and it has proved its excellence as an autonomous power, although one must naturally criticize the excessively unilateral results that have followed from this very autonomy, which too often hasn't conformed to the higher interests of the *entirety* of our economic order. And, no doubt, the most significant result of this entire period was the

---

[16] The Persian Emperor Darius, who ruled from 522 BCE to 486 BCE, suffered a crucial naval defeat at the hands of the Greeks at Salamine Island. In 1800, Napoleon won an important battle in Marengo, Italy.
[17] See "The Chief Features of the Revolution" in Arnold Toynbee's *The Industrial Revolution* (1881).
[18] French in original.
[19] Gian Pietro Carafa, aka Pope Paul IV (1476-1559). Latin in original.

identification of communism with the totalitarian order that reigns in Russia and, subsequently, with the perspectives of its partisans in our countries, who, over the years, have believed that Lenin and Stalin abolished capitalism. It pleases us to remember in this context that years before Karl Marx's *Grundrisse* was translated into Italian, our friend, the eminent economist Piero Sraffa, called our attention to the following passage in the book that settled the question: "To let salaried work continue and, at the same time, suppress capital, is an action that contradicts and destroys itself." Thus the social revolution that had been desired in the 19th century quite effectively became *utopian*, since it no longer existed anywhere in the global society where it might have been able to assert itself as what it could truly be.

Second, we have witnessed *the imposing reinforcement of the power of the States* as economic powers, political authorities and evermore refined organisms of surveillance. We can even say that, in this sense, the dream of the bourgeois economists of the 18th century (a legitimate dream, but one that often aroused the hostility of the aristocrats of the time) has been realized, but in a different form. The State theorized by these economists not only had to command the nation, but also to form and educate it in a specific way. According to Turgot, Quesnay, Letronne, Mercier de La Rivière and so many others, it was the task of the State to shape the spirit of its citizens according to a certain model that it proposed; the State must inculcate in them certain ideas and sentiments that it judged to be useful and necessary to overcome the obstacles that social reality presented to its activity. The economists of that period said that the State had to reform its political and civil institutions, and even the conditions of the lives of its citizens, so that they could be transformed. Bodeau summarized these ideas by advancing this prophecy, which was very radical for his times: "The State makes men as it wishes them to be."[20] In the 19th century, a very cultivated aristocrat, who was nevertheless too attached to the past, accused these economists of trying to create

> an immense social power that isn't merely greater than all those that currently exist; it is also different from them in its origin and character. It does not proceed directly from God; its origin doesn't lie in tradition; it is impersonal; it doesn't identify with the King, but with the State (...) This democratic despotism (abolishes) all hierarchies in society, all class distinctions, all fixed ranks;

---

[20] French in original.

composed of individuals who are almost identical and completely equal, this confused mass recognizes only one legitimate sovereign (the State), but it has been carefully deprived of all the faculties that could permit it to lead or even oversee its government.

The economists defended themselves against these accusations by invoking public education. Quesnay said, "despotism is impossible if the nation is enlightened."[21] The demands that these economists advanced were indeed better founded. Before the French Revolution, Letronne noted that, "for centuries, the nation has been governed by false principles; everything seems to have been done by chance."[22] Today we see what they foresaw. Perhaps it is fitting to emphasize that, a century before Marx, contemporaries of these economists, working in the same direction, advanced the current of thought that was subsequently called socialism. For example, one finds in Morelly's *Code de la Nature* all of socialism's doctrines concerning the necessity of reinforcing the power of the State, and in this work he foresees "the right to work, absolute equality, the uniformity of all, [and] mechanical regularity in all of the movements of individuals." It is surprising to see that in 1755, when Quesnay founded his school, Morelly recommended what is only today being fully realized everywhere. For example, we read in *Code de la Nature* that "the towns will be built according to the same plan; all the buildings used by individuals will be similar (...) Children will be removed from their families and educated in common in a uniform fashion, at the cost of the State."[23] The Statist centralization engineered by the bourgeoisie and the socialist bureaucrats was the product of the same necessity and the same terrain; each of these powers is, with respect to the other, like the cultivated fruit and the natural tree. But everywhere the State has become the protagonist that, with more or less efficiency, plans and programs the life of modern society. Therefore, the State is the *palladium* of market society, which converts even its enemies into property owners, as has happened in Russia and China, for example. And this fact allows us to remark that we do not fear resurrecting the old and noble term "market society." All of the grandeur of the world has been provided by merchants and the societies that they have built. Art, philosophy, knowledge in all its scientific and technical forms, political

---

[21] French in original.
[22] French in original.
[23] French in original.

freedom in its actually practicable modalities – all this only appeared in history, and has lasted, with the emergence and survival of the mercantile bourgeoisie and within the exact limits of its local or universal domination.

Third, *the isolation and the separation of people from each other has been highly perfected.*[24] Everything that could more or less directly disturb the tranquility of the social order, everything that could unite individual communities, corporate bodies, the neighborhoods of old towns or villages, and even the customary clienteles of cafés and churches, have been almost completely dissolved by the putting in place of the new conditions of everyday life and the new urbanistic countryside. We can say that each person now finds him- or herself in a direct relationship with the powerful center of the system that commands even the details of existence, and this center appears to each person, either successively or simultaneously, in its restrictive aspect as governmental authority, in the choices made by industrial production as to what will be available on the market, and in the selection of images to be contemplated. Thus the masses consume and watch what they want among the diverse things that are programmed for them, but they can only want what is available.

Fourth, *we have witnessed the unprecedented increase in the power of the economy and industry.* The modern economy has succeeded in giving a value and a price to everything, thus permitting everyone to consume the commodities that industry produces. We might even say that, to the extent that it has satisfied the essential needs of the population, the modern economy has been in the position to offer that population unnecessary things. Thereafter, what was inessential became necessary and this in the double sense that, subjectively, those things came to be perceived as such by the consumer and, objectively, they came to constitute a necessity for the industrial expansion that produced those precise commodities. Thus, at the moment that the citizen as consumer gained free access to the superfluous, all that was appreciated by the people of the past and all that was indispensible to guarantee them the maintenance of poorer and more precarious realities became *useless* and disappeared. From food to the entertainments of free time or vacations, there no longer exists anything that cannot be produced industrially, that is to say, cannot bring in an economic profit.

We do not want to deny that these developments also resulted in previously unknown inconveniences, such as new diseases caused by

---

[24] See "Separation Perfected" in Guy Debord's *The Society of the Spectacle* (1967).

pollution, etc. But, in any case, the very progress of science – the science of pharmaceuticals, for example – in its turn furnished antidotes that, industrially produced, constituted more commodities that could be sold to the population.

The system came to make use of (as an attribute of its sovereignty) the still growing distance between these rapidly changing realities and the words and feelings that now only correspond to appearances.

Popular notions, rooted in place for generations, no longer bear any relation with the completely different realities that have been produced by the most modern industries. Whether it is a question of what one used to call work, vacation, meat, influenza, or house, economic and Statist power makes use of all its elements to make known the modifications introduced into these realities. This power itself experiences modification, either by chance or by pursuing deliberate goals. And yet people *still speak of other things,* the things that have disappeared, using the same old words, which are also used during their debates on electoral programs.

Fifth and last – and this result concentrates together all the previous ones that we've enumerated – we have seen the vertiginously growing complications of the daily intervention of human society on all aspects of the production of life, and the replacement of all apparently natural elements by new factors that we could call artificial, fully justify the indivisible authority of every expert who builds or corrects the new economic and ecological equilibriums without which no one could live.

Therefore, there are now only experts in the workings of the State and the economy, because there are no operational fields or diplomas outside of these areas. And so *the existing hierarchy is forced to develop the secret and control in everything,* even when it doesn't want to do so. But all the hierarchies in history have always wanted to develop these things, even though doing so wasn't obviously necessary for everyone's interests. The double advantage that we derive from this situation resides in this: discontent with our society no longer makes sense, at the very moment that it has spread wider than ever before and concerns every single detail. Today, only total refusal, which is always difficult to formulate and put into practice, has a meaning that is threatening to our social order. And this threat is itself attenuated to the extent that a refusal of this kind, deprived of an exact comprehension of the totality and disinclined to envision the repercussions of real, historical confrontations, has the greatest chances of being stupid and contenting itself with some ideological illusion that leads its adherents astray.

Here, in brief, is how modern capitalism has been able to make the entire population participate in the freedom that it has built. And it is right to rejoice in this fact, because this enterprise had never been undertaken before, and bad omens piled up at the beginning. Perhaps a more lucid comprehension of history – for a century neglected in favor of economic studies that were themselves poorly disengaged intellectually from theology – would have inspired more confidence in the *elite*[25] of the time, who certainly could not have exactly foreseen the appearance of forms of domination that we have characterized here, but who could have speculated more boldly along the general line of the evolution to come, and thus perhaps more consciously hastened the useful formations? At the same time, one might have been spared a certain number of inconveniences from which we still suffer, such as the regressive mutation of capitalism in Russia. Let us reaffirm the point: despite the often legitimate, but many times exaggerated worries that the question has aroused in the dominant classes of almost all the countries, *capitalism must be democratic* because it can be nothing other. A glance at history, not to mention the most attentive and sharpest study of it, always leads us to the undeniable result that capitalism could never have grown, whatever the location, without a democratic society, [that is to say,] in the precise layer of society that lives the democratic life, wants it and needs it. And to deploy itself fully and completely, to transform everything into a commodity and incessantly renew the totality of commodities, capitalism must permanently give the entirety of the population a choice, the terms of which have been fixed by capitalism itself. Because one must be able to choose between two equivalent commodities, one must also be able to choose between two representatives. He who remembers fascism, who knows how badly State capitalism is managed by the totalitarian bureaucracies in the East, or who considers the permanent atrophy of the development of the merchant class in ancient Oriental despotism, will find the proof *a contrario* of this axiom.

Those who do not understand the necessity of remaining free quite simply do not have the good taste to do so, and we must give up trying to convince mediocre minds that have never known this sublime taste. The impassable limits that democratic freedom implies are its own safeguard, and it is reality that imposes them on it. Nevertheless, we can conclude that the peoples of the world have been more interested in concrete reforms put into action by democratic capitalism than in the multitude of sermons in favor of an abstract and total "freedom," a "freedom" that no one has ever seen

---

[25] French in original.

because it has never been realized. Thus freedom can only be understood on the basis of the actual reality of democracy, without being frightened or getting enthusiastic about the monotonous illusions that are always springing up about it.

No sensible person would think to deny the fact that, from its first admirable appearance in history, participation in the political management of democracy has been a domain reserved for a class of rich merchants or property owners, whether it was in the Athens of the 5th century [BCE] or the Florence of the 14th century. We see nothing different from this pattern in the famous year 1793 or anytime since then – beyond the fact that the dominant class of today isn't as well served by the always more numerous personnel to whom it has delegated the tasks of political administration, and nowhere as scandalously as in Italy where these roguish and incompetent domestic servants have allowed the roast to burn while they have nabbed the loose change from the pockets and drawers of their masters. As for the quite notorious other side of the democratic republics, we would like to say that the always-resurgent excesses of the infinite pretentions of the working classes quite clearly constitute the opposite of this democracy. The proof of this is that they have always resulted in immediate loss. But we are no longer at that moment in history when democracy – put into place or realized in a few cities – could have succumbed under the blows of these pretentions without impeding the general growth of a capitalism that was still generally sheltered in its previous social relations. Capitalism seized hold of the world for its own ends. The democratic order must be defended without any thought of retreat, "not only with the spear, but with the axe"[26] because, at the same moment that it is defeated, capitalism will definitively succumb, too.

We ask those minds and hearts that have become discouraged because, for the last ten years, they have taken the end of the troubles of a particular time for the end of the time of troubles, "Must we be resigned to the idea that any certainty that has been triumphantly conquered will be ceaselessly put into question, and is the crisis in society destined to always last?" We will respond coldly, "Yes." We must confront the harshest truth, "the truest cause" (to quote Thucydides) of this social war, which is unfortunately but unavoidably permanent. Our world *is not made for the workers,* nor for the other strata of impoverished salaried workers whom our reasoning must

---

[26] A quote from Herodotus, Chapter CXXXV of *The Histories.* It also appears in "Investigations without a Guidebook," an essay published in *Internationale Situationniste* #10, March 1966.

place in the simple category "proletarian." But every day our world must be made *by* them, under our command. This is the fundamental contradiction with which we must live. Even during the calmest days, the spark that could rekindle all of the masses' insatiable passions and their limitless and unstoppable hopes always exists in the cinders. This is why we never have the right to abstain from being intelligent for too long.

Truthful Report on the Last Chances to Save Capitalism in Italy

Chapter II:
How Capitalism Was Badly Managed in Italy and Why (1943-1967)

O my own Italy, though words are vain
The mortal wounds to close,
Unnumbered, that they beauteous bosom stain (...)
That your truth is understood here through my mouth – the one whom I could be.

Petrarch, *The Songbook*[27]

We have rapidly enumerated the objective successes that modern capitalism obtained prior to the last few decades. But since we do not intend to make an apology for this world – an apology whose utility in the proper domain of propaganda we do not deny –, we must set out in several summarized lines the origins of the internal crisis in our own country, a crisis that we are called upon to understand and confront without delay.

We know that, in the States, an illness is at first difficulty to recognize, then easy to cure, and that, through its progress, the disease becomes ever-more easy to recognize, but more difficult to treat.[28] As for what concerns

---

[27] The first three lines here are from Canto XVI, "To the Princes of Italy, Exhorting them to Set Her Free," but the concluding two lines are not. Perhaps they were taken from Pierre-Louis Ginguené's *Histoire littéraire d'Italie* (1819).

[28] Niccolo Machiavelli, translated by Angelo M. Codevilla, *The Prince* (Yale University Press, 1997), p. 11: "[B]ecause, by providing for oneself beforehand, one can remedy them easily, but if one waits until they draw close, the medicine is not on time, because the illness has become incurable [...] [I]n the beginning of its malignity, it is easy to cure and difficult to know, but in the progression of time, not having known it at the beginning, nor medicated it, it becomes easy to know and difficult to cure. So it happens in the things of state; because, knowing far-off (which is not given except to the prudent) the evils which are borne in it, one quickly cures them, but, not having known them, one allows them to grow so that anyone knows them, there is no longer any remedy for them."

Italy, we are convinced that, if we have so far been spared a pure, simple and irreversible politico-economic disaster, this has been thanks to the relative, contingent weakness of the adversary's forces and less so due to the merit and prudence of our politicians.

If we want to avoid a situation in which the illness becomes too easily recognized without relying on chance or hope, we must *immediately* diagnose it and simultaneously begin shock treatment before the workers understand its proportions and seriousness, which would inevitably open up to them new possibilities and new pretexts for struggle, as well as radiant perspectives of victory. The current wait-and-see attitude of the ruling class, which always fears to act or only acts out of fear, makes it look ridiculous in the eyes of the uneducated, working class masses. People are tired *for a while* before they perceive that they are, and nothing animates and supports a movement more than the ridicule of those against whom it is directed. Such situations are always very dangerous for both parties because they cause impotent despair in one and fatal fervor in the other. To not fall into the opposed risks of dramatizing or de-dramatizing the current crisis, there is only one route: to understand the nature and real depths of it exactly.

Our history from 1943 to 1967, when seen from a distance and in its entirety, appears to us as the representation of a fierce struggle that, in its first five years (up to the elections of 18 April 1948),[29] was seen in the majority of the countries opposed to the *Ancien Régime* of the Kingdom of Italy, which was born old and of which fascism was the supreme episode and the most recent archaism. It was exactly the Kingdom's traditional routines, its hardly glorious memories, its always disappointed illusions of grandeur and its mediocre representatives to which the entirety of the new Italian society was unanimously opposed, like a single person.

From the moment that the *Ancien Régime* was permanently destroyed, the elections of 1948 definitively concluded this first period of unified collaboration between the bourgeoisie and the lower classes of our country. By putting an end to the illusions of the workers, who still hoped for a possible collaboration between their parliamentary representatives and those of the wealthy classes, the bourgeoisie showed itself to be more realistic than the workers were. The triumph of the middle class was double: over all those who had been *above it* in the defunct Kingdom, and over all those who had been

---

[29] Thanks to financial assistance and clandestine "hit squads" provided by the CIA, the political right-wing won the Italian elections of 18 April 1948, which were "in danger" of being won by the Communists.

below it. This was a complete triumph, but it was only definitive in relation to those who were above the bourgeois, that is to say, the old decadent aristocracy of the large landowners. In this sense, the victory was effectively *complete* because all the economic and productive powers, and all the prerogatives and the government of the young Republic in its entirety were united as a monopoly within the boundaries that defined this bourgeoisie, which from then on became the unique leader of the ex-Kingdom. It took positions in all the useful posts of power by prodigiously multiplying their number, and very quickly got accustomed to living there, as much upon the public treasury as upon its own industry.

But this was, moreover, a *provisional* success because all the classes that had also contributed to the struggle against the Kingdom – first under fascism, then during the Resistance, and finally during the era of the Constituent Assembly – saw that the largest part of the fruits of victory were "expropriated" at the very moment when this victory became definitive. In such a situation, it wasn't a good thing to have too many illusions about the possibility of avoiding a new confrontation within the very interior of the heterogeneous coalition of the forces that emerged victorious from the preceding conflict. This conflict, which itself was part of a vaster conflict of global hostilities, had nevertheless quite weakened the working population and thus permitted the bourgeoisie to dedicate itself to its own interests without fear of once again finding itself obligated to measure up to a strong and unified adversary. On the other hand, after 1948, two decisive events contributed to once again reinforcing the position of the new dominant class: above all, the political strategy chosen by Togliatti[30] for the Communists and by the Left in general was not at all in contradiction with the new needs of the democratic and liberal center since, under the sufficiently vague mandate of the economic "reconstruction" of the country, renewed social tensions were momentarily frozen and, reciprocally – to the extent that this reconstruction was effectively undertaken – political passions calmed down and a public and private wealth such as Italy had never before known developed very rapidly. No one can forget how the Cold War, which excessively augmented international tensions, opportunely served to cool and defuse the real reasons for the internal conflict, which was constantly projected beyond Italy's frontiers. The insurrectional episode of July, 1948, for which the attack against Togliatti served as a pretext, was the only noisy consequence of the workers'

---

[30] Palmiro Togliatti was head of the Italian Communist Party until his death in 1964.

disappointment after the elections of 18 April, and this was the occasion on which the Italian Communists, who loyally repressed the insurrection from within, with their own troops, proved their coherence and their responsibility with respect to their democratic political choices.

From then on, the particular needs of the bourgeoisie became the general needs of the republican government. They also dominated both the foreign policy and the domestic affairs of the country. The spirit of the times was active, industrious, poised; what one calls political dishonesty had precise justifications; it was, by temperament, a timid spirit, but was rash due to egotism, and moderate in everything except its mediocre taste for "well being." This spirit would have accomplished miracles if only it had possessed a little of the nobility of intention that has always appeared indispensible to us, but, by itself, this spirit could produce nothing other than a series of weak governments, without virtue or grandeur. Master of everything as no other aristocracy on the peninsula had ever been, the middle class or, rather, that part of this class that we could call the class of government, had taken up its residence in governmental power and, soon after, in its idiosyncrasies: the government took on the appearance of a private industry and was no longer the political expression of private industry properly speaking. None of the members of this class appeared to think about public affairs, not even to make them turn a profit for their own private interests or their own political current, while the holders of economic power and the common people – in a blithe thoughtlessness that united them for a while – occupied themselves with their respective individual interests, which were great in the case of the former and small in the case of the latter, with both contributing to the deceptive success of the ideology of well being.

Posterity, which only sees the brilliant crimes and ordinarily misses the vices that are at the origins of all the most serious crises, will perhaps never know how all the successive Italian governments had gradually but increasingly taken on the appearance of a commercial firm in which all the operations were made in view of earnings that could be derived for its particular associates, naturally under the sign of the public interest. When some of the most authorized representatives of economic power began to worry about the risks and the *costs* of a parallel system of government, the leaders of Christian Democracy, then accustomed to consider any government ministry as sinecure guaranteed to each of its notables, resorted to the saddest kind of blackmail by threatening to render public several virtual scandals in which economic power wasn't any less implicated than political power, with the intent of keeping the reins of the government locked into *imbroglio* and bankruptcy. It was certainly an error to give in to this blackmail. Almost all of the political despicable acts of which we have been the unwilling and mostly powerless witnesses have, in our country, followed from the fact that the men who are introduced into political life – deprived of a personal inheritance – fear their ruin if they abandon their places in government or from the fact that their ambitions, personal passions or fears render them so obstinate in the continuation of their careers in power that they consider the simple idea of abandoning them with a kind of horror, which distorts their judgment and makes them sacrifice the future to the benefit of the present and their honor to the roles that they play.

On the other hand, no one can forget the responsibility of America, which seems to have accorded more confidence to the forced and artificial stability of the Italian political class – which obviously presented as its own work the recent well-being to which the country had acceded – than to the real craftsmen of the economic miracle, who were the industrialists and entrepreneurs, in general.

The current politico-economic paralysis, which had to be the direct and principal result of such irresponsible conduct, was the least unforeseeable thing in the world and yet it was regarded as a Cassandra-like prophecy that could have warned against such a possibility, which was what we exhausted ourselves trying to do. If our efforts weren't publicly mocked, this was, in the best of cases, due to a residue of respect and, most often, due to pure and simple fear. Instead of praises for our alleged foresight, which at the moment come to us from all sides, we would have more modestly preferred a more attentive audience at the moment when there was still time to avoid this dreadful situation.

In a political world composed and led in such a fashion, what was most lacking was political life itself. On their side, the majority of the industrialists and, more generally, the holders of economic power, who were once again too devoted to their religion of *laissez faire*,[31] didn't entertain with sufficient clarity the consequences (obviously more damaging to them than to the politicians) of such a doctrine when it was set up as the unique rule for Italian politics and were too trusting of an inertial power that had made the politico-economic machine, following its own internal rules, function "automatically," and all the more so when one kept one's hands off of its delicate mechanisms. What one cheerfully forgot were the very society in which this "automatism" functioned and the profound transformations that it had brought about over

---

[31] French in original.

the prior 20 years. The industrialists, who were rightly bored by the empty and verbose speeches of the government, placed, on the other hand, an extravagant confidence in the simplistic technical studies made by mediocre economists with whom they surrounded themselves and from whom they asked forecasts that reassured them concerning the expansion of and increase in their profits. With the arrival of the critical moment in which these forecasts were challenged point by point by the facts, *the industrialists asked for more forecasts,* as if to compensate for real losses with illusory certitudes, to which they hastened to make themselves slaves. A collective neurosis seemed to have seized these men, the majority of whom lacked the mental strength of their fathers and the character traits of their ancestors. They had inherited their money but not their courage, their pride but not their dignified prudence. The first failures sufficed to depress them psychologically and to remove from them the spirit of free enterprise. Thus they progressively lost the indispensible class solidarity that should have been their first line of defense when they were confronted by the excessive political power and the growing pretentions of their workers – and all this deteriorated into a kind of law of silence; they became accomplices in a shared impotence with the political class that, in truth, they allowed themselves to be fleeced by.

The nation in its entirety then overtly felt a tranquil contempt, as much for economic power as for the political administration, and those concerned were quite wrong to consider this tranquility to be confident and satisfied submission, the forthcoming end of which they did not perceive. Slowly the country divided into two unequal but still not opposed parts: on high there reigned apathy, boredom, impotence and immobility; down below, by contrast, political life began to manifest itself in feverish, irregular and *apparently* extra-political or extra-unionist symptoms that an attentive observer could have picked out without difficulty. We have had the misfortune of being one of those observers, and consequently we were much more sensitive to the inquietude that grew and rooted itself in the heart of our society to the extent that public morals deteriorated into general indifference; we were no doubt favored by our personal integrity, which has always been above party interests, and by the fact that our interests have never been dependent upon chance. In addition, we were favored by our position, which has required a character hardly inclined to false fears and false consolations, and so it was easy for us to enter into the game played by these institutions, as well as the mass of small, everyday facts, where in complete coldness we examined the evolution of the morals and opinions of the country, among the ruling class as well as among workers. It was thus, and not at all thanks to the

Truthful Report on the Last Chances to Save Capitalism in Italy

chimerical wisdom that today one wants to attribute to us, that we have been able to clearly discern the many indicators that have ordinarily appeared in history in advance of each of its catastrophes and that always herald revolution.

Towards the end of 1967, these symptoms became so numerous that we believed it our duty to communicate in a private manner our preoccupations to the man who, due to the very position that he occupied, had to be able to understand (more than anyone else) the seriousness of the disastrous consequences, and who had the greatest interest in preventing them.

We then said to him that the Constitution of the Italian Republic had abolished all the secular privileges and destroyed all the protected rights, yet let a fundamental one (the right to own private property) continue to exist amidst the utopian perspective of extending that right to everyone. We then added that, in a period when half the States in Europe were confronting a growing discontent among the workers and the entirety of the young generation, the property owners shouldn't have too many illusions about the solidity of their situation, nor should they imagine that the right to own private property would continue to remain an insurmountable wall for the simple reason that in Europe, until then, it had never been breached, *because our times resemble no other.* We have shown how, at the origin, when the right to own private property was the only foundation required for the support of many other rights, we defended it without too many difficulties or, rather, our enemies didn't dare attack it directly. The right to own private property constituted a kind of wall within the wall of society, and all the other rights and privileges were its forward defenses. Blows could not reach it and, on the other hand, our enemies did not seriously seek to besiege it. But today, for many people the right to own private property seems to be the last remains of an aristocratic world that was destroyed *de jure et de facto.* Standing alone, it appears with the greatest obviousness to be a unique, isolated privilege in a leveled society, while all the other protected rights (much more contestable and justly hated) no longer serve as a screen, and so the right to own private property itself has been challenged in the most dangerous manner and with a contagious violence. It is no longer the attacker, but the defender, who seems obligated to justify himself.

Confirming our preoccupations and aggravating them with the stamp of an event, what took place in May 1968 showed the world that the time had come when our form of society was revealed to be divided into two large parties in the most unhealthy way. *Real* political struggle, which we could neither prevent nor win with speeches and which unavoidably had its theatre of operations in the factories and streets, henceforth broke out between those who possessed and those who were deprived of this right and, under a thousand diverse pretexts, our enemies did not miss an occasion to choose private property as the battlefield, and everyday and everywhere salaried work became a *casus belli*. Our political calendar could have been illustrated by an old maxim: "the illness never ends when those who command have lost their sense of shame, because that is exactly the moment when those who used to obey lose their respect for them, and it is at that very moment that they leave their lethargy behind through convulsions."[32]

Thus in France in 1968 and Italy in 1969, we saw our class tremble, without either courage or dignity, as if overwhelmed by the phantasm of its imminent death. Subsequently, this very bourgeoisie, as if awoken from a nightmare, believed itself to be definitively saved, but without seeking any further explanations. We never allowed ourselves to share either one of these errors, because we still heed the effects that passing whims, determined by this or that circumstance, can have on the human spirit, and because we are too well informed about the singular doctrines that, from time to time, appear or are rediscovered everywhere and that, under different names and labels, have had as their common denominator the denial of the right to own private property and the contestation of the duty of salaried work. The seriousness of the situation in which these things came about could be measured by the extreme ease with which these ideas spread in the factories, neighborhoods, schools, and offices, and the enthusiasms that they aroused.

"Beauty," Stendhal says, "is the promise of happiness,"[33] and we acknowledge that all the new theories, and the ideas that have simply been sketched out, denounce above all the pallor, boredom, and *routine*[34] of everyday survival in industrial societies; the real ugliness that has overcome the appearance of our towns that have been abandoned to the ravages of urbanists and speculators of all kinds; the pollution of the air, food and minds that has been democratically imposed on all the inhabitants of the urban centers. As a result, we easily understand that this "global" critique, even if it is generally imprecise, has easily hit the bull's eye for people who are bored and impatient with the so-called diversions and *leisure activities*[35] that this society can offer them, and we can likewise explain how at present it has

---

[32] French in original: Cardinal de Retz, *Mémoires*.
[33] French in original.
[34] French in original.
[35] French in original.

Truthful Report on the Last Chances to Save Capitalism in Italy

become objectively easy to make the workers believe anything that comes from channels of information that are different from the customary ones, which are accused – often rightfully so – of hiding the truth and being specialized in the manipulation of lies in which the majority of the country has believed for many years. Disappointment, the effects of which are always dangerous, seized the petit-bourgeoisie, which in these last few years has seen the disappearance of the social promotions that had been promised to it by the political parties that it voted for. The disappointment of the petit-bourgeoisie, which we should fear less than the rage of the workers, first manifested itself through the contestation that the children of this class engaged in at the high schools and universities, and subsequently it spread to their families, who were politically oriented toward the right-wing opposition parties or, in the majority of cases, the left-wing ones. The Communist Party was therefore able to offset the electoral losses that had cost it the defection of a part of its base among the workers, who became radicalized and escaped from its control. But what appears to us the most immediately worrisome development is the vulnerability to illusions of happiness and beauty that our political class has created in all the classes that, due to vocation or disappointment, are now openly opposed to the bourgeoisie, which has prepared the battlefield without preparing itself for battle against the other class, thus forgetting the following infernal prophecy.

> For all eternity they will be against each other:
> The one lot will arise out of their graves
> With fists clenched, the other with their hair cut off.[36]

---

[36] Dante, *The Inferno*, Canto VII, lines 55-57.

Gianfranco Sanguinetti

Chapter III:
In Which the Social War Begins Again, and Why Nothing is More Disastrous than Believing that It was Won
(1968-1969)

"What causes apathy in the States that suffer from it is the duration of the illness, which seizes the imagination of men and makes them believe that it will never end. As soon as the day comes when it does, which never fails to happen when the apathy reaches a certain point, they are so surprised, so relived and so carried away that they immediately swing to the other extreme and, although they are far from considering revolution to be impossible, they believe it to be easy, and this disposition is sometimes capable of making one on its own."

Cardinal de Retz, *Mémoires*

Our social preoccupations were obviously not born from a romantic outburst of the heart, but intelligent reflection, because in the relative but incontestable poverty of certain social strata, we don't see suffering that must be cured – a demagogic utopia on which we will willingly let others speculate – but *a disorder to be prevented.* Yet in no other period of history have so many principles and concepts been enunciated, and with so much pretense and claims to universality, where this matter is concerned. If history seems to most often present itself as a conflict of interests and passions, our recent history up to these last few years – although passions have not been lacking – has mostly presented itself, instead, as a struggle between *principles of justification* and partly as a struggle between subjective passions and objective interests that are almost always hidden behind the flag of "superior" justifications.

Over the years, we have impassively witnessed the lamentable spectacle presented by our bourgeoisie, which has justified itself to the other classes by what it intended to do in defense of the "exploited" people and, reciprocally, the other classes, which work at this project all the time, were accused of pursuing egotistic interests. This was one way among others – although a less than useful one – of passing the time at a time when one could still allow

Truthful Report on the Last Chances to Save Capitalism in Italy

oneself to waste it. For our part, we note that the quite respectable and artificial interest of these gentlemen in social questions had a principally psychological origin. This interest was its own justification, and more or less responded to the "moral" need to soothe one's conscience in one manner or another during the period of the "economic miracle," which made these men quite euphoric. With an academic casualness and a studied ignorance, they discoursed about social questions, because the new middle class believed them to be nearly resolved and hadn't known about nor comprehended the magnitude of the revolutionary jolts of 1919-1920, nor even how the bourgeoisie had defeated them. However, in reality, a solidly unified and vague worry about, and genuine disinterest in, civil society was hidden behind this "sensitive" façade. Among the members of the bourgeoisie, class spirit had been lost, and this corresponded to the loss of its self-assurance and the acquisition of a great timidity. In our opinion, this new bourgeoisie feared being right and feared being afraid. Shortly thereafter, they came to realize *that they were right to be afraid.*

The ruling class's lack of interest in the mutations then taking place in civil society reached its height when an unforeseen fact of global scope was suddenly revealed, but in a traumatic way.

The insurrectionary events that shook France in May 1968 unquestionably showed that a new social revolution, one unburdened of all previous illusions and delusions, was knocking at the door of modern society. At first it wasn't understood and then it was hidden – not without reason – but this insurrection was, due to its very existence, the most scandalous and terrible failure that the European bourgeoisie had suffered since 1848. As in 1848, the wind of revolt blew all over Europe, and it was inhaled in France as in Germany, in Italy as in Czechoslovakia, in Yugoslavia as in England. In different forms and diverse fashions, the thoughts and actions of the populations in open revolt against society turned against the world that is ours, and these were the same populations that (no less than the ruling class) seemed to have forgotten for a half-century what people in the 19th century called the "social question."

We need not insist upon recalling here that, in 1968, France experienced the most extensive and longest general strike that had ever paralyzed the economy of an advanced industrial country, and that this strike was also the first "spontaneous" general strike in history. For several weeks, all of the powers of the State, the political parties and the unions were quite simply *effaced,* and the factories and public buildings in all the cities were occupied. Because we do not want to obligate anyone to share this opinion, it is outside the scope of this pamphlet to demonstrate why the events of May were profoundly revolutionary and virtually much more dangerous to the world than the Russian Revolution of 1917. Thus we will limit ourselves to considering *the facts* that these events set a very menacing precedent and that the ideas of the movement that began then and there have spread everywhere, because everywhere in Europe the poor classes have grown in number, their importance has grown more than their way of life, and their aspirations have grown more than their power.

Ever since the French Revolution of 1789, that is to say, ever since the bourgeoisie seized hold of the political responsibility for the management of the States all over Europe, the people in these countries have sought to throw off their conditions, thus periodically changing all of the political institutions. But after each change, they have discovered that their lot hasn't truly improved or that it has been improved with an unacceptable slowness with respect to the speed of their desires. Thus it was unavoidable that, one day or another, the workers would finally discover that what has confined them in their situation wasn't the *constitution* of the different States – kingdoms or republics, fascist or Socialist dictatorships, parliamentary or presidential democracies – but the very laws and principles that constitute all modern societies, and thus it was natural that the poor classes sooner or later came to wonder if they didn't have the power – and perhaps the right, as well – to change those laws as they had changed other things. And to speak specifically of private property and the State, which are the foundations of the entire social order, wasn't it an unavoidable consequence that they were once again (but in a completely new way) denounced as the principal obstacles to the demand for equality among men and women, and that the idea of abolishing them completely – and not in the manner that one once said they had been abolished in Russia – came to the minds of all those who felt that they were subjected to and excluded from them?

This natural inquietude in the spirit of the people, this unavoidable agitation of their desires, this resentment of unfulfilled needs, and these mob instincts formed, as it were, the fabric out of which professional agitators wove monstrous or grotesque figures, which were rejected by all the political parties and especially by the Communists. In May, in Paris, each person proposed his or her own plan for the construction of the "new society." One demanded the immediate abolition of salaried work; another the inequality of the distribution of goods; a third wanted the end of market society and the oldest of the inequalities, the one between men and women; all seemed to agree to exclude all kinds of external authority, to experiment with forms of

Truthful Report on the Last Chances to Save Capitalism in Italy

direct democracy, to reject all institutions, political parties and unions.[37]

The most attentive observer was struck by the fact that, quite contrary to what was collectively said at the time, the overwhelming majority of this movement wasn't composed of students, but workers and other salaried employees. One could obviously find utopian or simply ridiculous ideas among them, but the terrain on which these ideas were nourished and propagated is the most serious subject that the political parties and statesmen can examine today, because what is in question is our very world.

In France and Czechoslovakia, where this insurrectionary movement (it would be more exact to call it a revolutionary movement) had principally taken hold, who repressed it with the greatest efficiency? Who favored or imposed the return to normal in the factories and streets? Well! In one case as in the other, it was the Communists: in Paris thanks to the unions; and in Prague thanks to the Red Army. This is the first lesson that we can draw from those events.

But the social sickness that produced the most conspicuous symptoms in France was quickly transformed into an epidemic, and Italy was subjected to the contagion in a completely unique way. The incubation period and the development of the sickness came so close together in time that here it is a question of writing history, and that history is still so well engraved in our memories that it would be useful to retrace it in this pamphlet. It is sufficient to remember that the so-called student protests were naturally, here as elsewhere, ephemeral and quickly became a simple phenomenon of depravity – tolerable due to the presence of so many others – that occupied the pages of the daily newspapers and the discourses of the intellectuals rather than a vital sector of productive society. Nevertheless, each person knew that a quicker, less apparent but much more worrisome movement – parallel to and contemporaneous with the student movement – had begun in the factories, at first without connecting links or widespread publicity. Despite the traditional unionized management of the Italian working class, Italy also saw its first forms of "spontaneous" struggles and para-union strikes. Precisely because the significance of this phenomenon was underestimated at the time, it was easy for it to spread during the following months with a growing radicalism. A kind of frenzy seemed to have seized our workers who, united into so-called "base committees," began in an autonomous manner to advance extravagant extra-salary claims that were sometimes colorful and sometimes absurd, but

---

[37] Direct quotes or paraphrases from Alexis de Tocqueville, *Recollections of the French Revolution of 1848*, published 1893, posthumously.

always noxious because, in every case, they found partisans who were ready to fight for them. Leaving aside all the other examples, we will mention the one furnished by the employees of an important public enterprise in Milan, where at the end of 1968 a "base committee" organized (and with "success") a series of strikes that aimed at getting the time it took the workers to get from home to their workplace counted as time at work and thus subject to compensation as such!

We had the impression that the workers were literally in competition to see who could record the greatest amount of damage with their disastrous fantasies. In reality, the declared goal of each particular conflict was out of proportion with the social damage that the generalization of the strikes and demonstrations of all types caused to the country. In our opinion, the rest of the workers did not care what they combated: what they wanted was combat *itself.* Thousands of pretexts were found, but this was the single undeclared goal, and no salary increase would suffice to appease them.

We know that it was, nevertheless, only in 1969 that Italy experienced all of the fateful "modernity" of its social crisis. In fact, it was the first serious disorders in the prisons and factories of the North, along with the revolt in Battipaglia in the spring of that year,[38] that illustrated the extension of the crisis from one end to the other of the peninsula and that could be called the "qualitative leap" of the crisis' seriousness with respect to the prior year. In truth, the passions of the students of 1968, despite their claims that they were from "the Left," didn't go beyond politics, while the passions of the working class were *social,* and our readers will not be ignorant of what this inevitably implies. The workers did not ask for this or that reform; they did not contest a policy, this government or that government, or one political party or another, but society itself and the bases upon which it rests.

And yet, despite all this, we can affirm that in this period the government was not as alarmed by what took place in the country as were the leaders of the Communist opposition. In the first phase of 1969, the only people really and truly worried about the near future were a few union *leaders*[39] and officials of the Communist Party, because they were the only ones to observe the working classes from close range, each day registering their mood and subversive will. The state of permanent agitation in the country had already surpassed not only the hopes but also the desires of the

---

[38] The revolt occurred on 9 April 1969 in response to the closing of a tobacco plant, which was one of the biggest employers in the region.
[39] English in original.

Truthful Report on the Last Chances to Save Capitalism in Italy

most fervent unionists, that is to say, those who believed (wrongly) that they were at the origin of the phenomenon. This wasn't the first or the last occasion in which we were able to recognize the lucidity of the Honorable Giorgio Amendola,[40] but perhaps on this occasion he surprised us even more than usual and, as a result, we held him in even greater esteem than before. Unlike so many others, this politician possessed an agile spirit, cold but cordial, eminently subtle, which immediately went to the heart of any question, but didn't neglect the details, without prejudice and without rancor, a true connoisseur of the range of human weaknesses and penchants, especially where his party was concerned, and always capable of playing upon them when his interests weren't opposed to him doing so. In sum, he was a man whom we could not prevent ourselves from esteeming and listening to. And so much more so in such an epoch as post-1968 Italy, when the Honorable Rumor, President of the Council,[41] did not enjoy our confidence because he said things of this kind: "Be tranquil, everything will end well, there isn't a free government that couldn't surmount tests of this sort." We, who are less worried about the fate of the government than we are about all the other problems, we found that this response perfectly captured this resolute but limited man, limited with much spirit, but this spirit is of such a kind that – seeing clearly in detail all that is on his horizon – doesn't imagine that this horizon could change without warning. On the other hand, we must keep in mind the industrialists, some of whom – victims of an anguish that is confined to cases of pure and simple stupidity – imagined doing nothing more than calling the unions to order, as if the unions, from the moment that they weren't responsible for this situation, had been in a position to be officially opposed to it without running the risk of having the movement eliminate them and, this time, formally.

It was around the middle of 1969 that we came to explicitly demand from the Italian Communist Party [ICP] what guarantees it could offer the government to help it stop the workers' movement before autumn and what it would demand in return. The Communists, who knew better than anyone else the magnitude of the stakes and the danger of this movement, transmitted

---

[40] A member of the Italian Communist Party, Amendola (1907-1980) favored non-Marxist moderation in the Party's dealings with the government and the economy.

[41] Mariano Rumor (1915-1990), a member of the Christian Democratic Party. In 1969, he was the Prime Minister of Italy and, in 1975, the Italian Minister for Foreign Affairs.

their wishes, but both political power and a large number of industrialists – either because they underestimated the risks of the months to come or because they overestimated the "risks" of any agreement with the ICP – found the compensations demanded by the Communists to be out of proportion to the guarantees that they could offer. With *a posteriori* knowledge, we can say that the Christian Democrats still ignored the strength and utility of a Communist party in such circumstances and that the ICP, for its part, underestimated the strength that the wave of "spontaneous" strikes would have in the following months, because the Communists counted on time and the "natural" speed of the events with a little too much casualness, awaiting the moment when they would be called, and the Christian Democrats counted too much on the fact that the Communists – so as to not come to an open break – had in any case to do what they had promised to do, even without receiving immediate compensation for it. The calculations of both groups would have been justified or justifiable if confronting a *political crisis* was the order of the day. Both sets of calculations proved to be insufficient, not to mention thoughtless, because everyone seemed to forget that Italy was actually in the midst of a pre-insurrectionary *social crisis*. From the moment that the Communist leaders, expecting subsequent developments, remained entrenched in a position that was no less rigid than that of the Christian Democrats, who nevertheless bore the initial responsibility for this stiffening and did so from the moment it became clear that, in this case, one could not come to the end of anything by this route and that one had to act immediately but in another way. What, consequently, was the direction to follow? We will answer with the words of a journalist (Nicola Adelfi, writing in the pages of *Epoca*), because a great philosopher who taught more than a century and a half ago pointed out that, "there is all of the truth and all of the false in public opinion," and because journalists are specialists in public and private opinions. To wit:

> A number of political, union-related and political symptoms make one think that this situation will continue (…) We don't see how the wave of violence can be broken or even simply attenuated. At least not without the occurrence of something unforeseeable and traumatic in nature: that is to say, something that, unexpectedly, deeply shakes public opinion and gives it the feeling of finding itself henceforth a step away from anarchy and its inseparable companion, dictatorship.

Truthful Report on the Last Chances to Save Capitalism in Italy

We couldn't have said it any better ourselves, but for something "unforeseeable and traumatic in nature" to take place, one needed to have, above all else, a homogenous and less fragile government than the Rumor-Nenni Center-Left coalition. We know that, after the formation of the first Center-Left coalition, various representatives of economic power took up or placed certain men in eminent positions in the unfortunate Socialist parties, which were called unified at the time. Well! To topple the Rumor-Nenni Center-Left coalition, it was enough, at the beginning of July [1969], to ask the Social Democrats (always ready to undertake operations of this kind) to provoke a new split. The unification intended to last 10 years collapsed after only 10 months. The next day, the government fell and, a month later, at the beginning of August, Rumor constituted his second "mono-color" government, in which all the currents in the Christian Democratic Party were represented, if our memory serves us well. Despite all of its inadequacies, Rumor's cabinet appeared to us to be among the most *efficient* in the history of the Republic, if only for the actions successfully executed by the Minister of Labor, the Honorable Donat-Cattin, and the Minister of the Interior, the Honorable Restivo, during the autumn of 1969, which since then – in an admirable *understatement*[42] – has been called "hot."

As the foreign press affirmed at the time, the only institutions that continued to function in Italy were the unions and the police, that is to say, the Ministries of Labor and the Interior. Carlo Donat-Cattin had in fact once been a union leader, and Franco Restivo, close friend of Vicari, then the Prefect of Police, had already had (with Vicari) experience with political terrorism in Sicily (of which Restivo had been the president) after the Second World War, when the bandit Giuliano ran wild.[43] Precisely in 1968, a number of small attacks using explosives – though they didn't have serious consequences – contributed to increasing the disorder that the protests by students and workers continued to create in the large towns, and even in the small ones. These were acts of narrowly limited scope in comparison, for example, with the acts of sabotage that were taking place in the factories. These limited attacks bore the signatures of fascist or Maoist groupuscules that there fighting their local adversaries, but these attacks were at the origins of larger ones and, as Tacitus says, "it will not be useless to study those things that, at first sight, are trifling events, because out of them the movements of vast

---

[42] English in original.
[43] Salvatore Giuliano had been the leader of the Voluntary Army for the Independence of Sicily. He was murdered in 1950.

changes can arise."[44] Because in Italy, at that time and afterwards, the unions and the police weren't the only forces that still functioned. For several months, the secret services had been silently at work, too. And since the political sphere continued to shilly-shally in the face of the worsening crisis, it was necessary to finalize (before the summer) a tactical diversion, an artificial tension of which the principal goal was to *momentarily* distract public opinion from the real tensions that were tearing the country apart. In the next chapter, we will see what were the undeniable advantages of such a tactic, and what were also the damages that it inflicted when it was transformed into a strategy, and we will therein render public the critiques that, in another place and at another time, we addressed to our secret services, which – due to a blunder that had no precedent in history – today are publicly exposed to the accusations of the first judge to come along and the entire country.[45]

And so, although the aforementioned small attacks were the *background*[46] for these tactical diversions, their proper beginning coincided with what took place in Milan on 25 April 1969 and during the month of August [1969]. The operations to which we have alluded here were, in a certain sense, a repetitive preview of the events that took place in the autumn of 1969. These events were not expected and, starting in September, the first acts of sabotage of considerable magnitude took place at the FIAT factory in Turin, the Pirelli factory in Milan, and hundreds of other places. The top-level negotiations concerning the renewal of the contracts between employers and unions were only one set of pretexts among many others. A number of actions and events – in a period that truly didn't lack them – were eclipsed by others that followed them in an always rising *crescendo,* and we can be dispassionate about them here because the profound meaning that this class war unwittingly[47] gave

---

[44] *The Annals,* Book 4, paragraph 32. Latin in original.
[45] In September 1974, General Vito Miceli, the head of the *Servizio Informazioni Difesa* (the Defense Intelligence Service), was arrested and charged with involvement in a failed coup attempted in 1970 by the veteran Fascist Valerio Borghese and Stefano delle Chiaie's neo-Nazi *Avanguardia Nazionale* organization. During his subsequent trial, Miceli defended himself by disclosing the existence of a "parallel SID" that had been formed as a result of a secret agreement with the United States within the framework of NATO (i.e. "Operation Gladio").
[46] English in original.
[47] The word employed here, *inconsciemment,* also means unconsciously and thoughtlessly.

Truthful Report on the Last Chances to Save Capitalism in Italy

itself through its intensive and extensive development became more important than any of its particular episodes, which were only the Roman mile markers along the road that led, always more obviously, to a social revolution.

In the course of our life, we have associated with well-informed people who have written history without getting mixed up in it, and we have had to act in concert with politicians who have constantly and uniquely been involved in the production and prevention of historical events without thinking too much about describing them in writing. We have always observed that the former see general causes everywhere, while the latter – living in the midst of everyday occurrences, which apparently produce each other – gladly represent things in such a way that all the events that serve them well must be attributed to their own personal merits, as if it fell to them exclusively to determine the course of the world, and as if any setback was only the consequence of this or that particular and absolutely unforeseeable event. There are times when both the historians and the manipulators of events are wrong and, if in this epoch one must expect everything, because everything is possible, we must not allow ourselves to be taken by surprise. For example, in the autumn of 1969, which Raffaele Mattioli[48] defined, with the philosophical detachment that was unique to him, as "the lyrical expression of history in action, where no one had the courage to be what he was," we witnessed the pitiful spectacle of industrialists placing more confidence in the unions than in themselves, and the unions placing their confidence in the concessions that they could obtain from the government, and the government placing its confidence in the efficaciousness of its special services. We were among a small number who knew that the worst that one foresaw was in fact *too optimistic,* just as today few know that Italy once more finds itself *only an hour away* from a general insurrection, and that if this has, fortunately, not happened yet, we have to thank the precautions taken by this or that person, and not the interplay of other factors.

The struggles surrounding the contract negotiations obtained notable success on the terrain of salary increases, but it was a pitiful illusion to believe that things would calm down once the new contracts were put into place. As we have already said, from the moment that the workers no longer fought to simply obtain salary increases, it was clear thereafter that, though such increases were constant, we could no longer hope to purchase social peace

---

[48] Raffaele Mattioli (1895-1973) was an Italian economist, banker and business executive. Censor's book is dedicated to him (p. 2).

with them. Such peace risked being no more than a happy memory of past times. In fact, when certain categories of laborers – such as municipal workers – obtained a new contract, they continued their illegal strikes under the pretext of supporting the struggles of workers in the private industries, for whom the negotiations remained suspended. On their part, the unions could not expose themselves to the danger of cutting themselves off from the working masses by disavowing all the strikes that the unions did not want to undertake and had not been able to prevent. On the contrary, they had to accept the existence of those strikes so as to not exclude in advance the possibility of being accepted by them in turn, at a later stage, as the authorized spokesmen for the workers' demands. To prevent open riots, the union confederations had to find other objectives than salary demands and then try to channel the workers' protests towards them.

It was in fact one of those objectives, which appeared artificial to the workers themselves, that furnished the occasion to unleash a blatant and obvious insurrection. On 19 November 1969, the unions announced a national day of general strikes over the question of rent. In Milan, this strike, which saw the largest abstention from work in the history of the Republic, degenerated into a riot very quickly. The union *leaders*,[49] who made speeches at the Lyric Theatre, were boycotted and insulted by the workers who, abandoning the meeting, severely attacked the forces from the Department of Public Safety, who were forced to withdraw from the entire neighborhood, and then the workers erected barricades in the center of the town.

We have precise memories of this spectacle, because around noon on 19 November we had to cross the via Larga to go to the home of an industrialist (not far from the location of the confrontations), where we were invited to have lunch with several politicians and other people from the economic world. Since it was impossible to find a taxi, we crossed a part of town on foot. We found the majority of the streets to be tranquil and almost deserted, as happens in Milan every Sunday morning in early hours, when the rich are still asleep and the poor are not at work. Here and there, from time to time, a young man – looking more like a suburban salaried worker than a student – tranquilly posted a placard on the façades of the buildings. He offered us several of them, signed by some group of "autonomous workers" or by a "base committee," and one of those manifestoes surprised us with its gloomy title, which was redolent of the 19th century and went something like this: "Notice to the Proletariat on the Current Occasions for Social Revolution."[50] Having

---

[49] English in original.

Truthful Report on the Last Chances to Save Capitalism in Italy

passed through the obstructions of the police and the demonstrators (not without some difficulty), we finally reached the apartment of our host, who was more upset than usual. The food was magnificent, as was customary, but the table was deserted. Of the half-dozen people invited, only one other person was present, and he was late and wasn't even the most eagerly expected guest. We sat with a passive air among this useless abundance, and a profound silence descended upon us after I made the simple observation that we live in strange times, in which, as Tocqueville noted in 1848, one can never be sure a revolution won't break out between the moment when one sits down at the table and when the meal is served.[51]

Telephone calls that relayed the news rendered the expectation of dire events even more unnerving. The news accumulated: a Public Safety officer was killed in front of the Lyric Theatre, and neither the police nor the unions were in a position to control the battlefield, which they had abandoned. All through the afternoon, the telephone line was the only umbilical cord that tied us to the world. The worst fears concerned the situation in Turin, because if the workers in Milan believed that the situation there had also escaped from our control, the *chances*[52] that the riot and the strike would remain limited to

---

[50] Cf. *Avviso al proletario italiano sulle possibilita presenti della rivoluzione sociale* ("Notice to the Italian Proletariat on the Current Possibilities for Social Revolution"), a tract written and distributed on 19 November 1969 by the Italian section of the Situationist International, of which Gianfranco Sanguinetti himself was a member.

[51] Cf. the skit by Monty Python's Flying Circus entitled "Party Hints" (1972), in which "Veronica" gives the following advice. "This week I'm going to tell you what to do if there is an armed Communist uprising near your home when you're having a party. Well, obviously, it'll depend how far you've got with your party when the signal for Red Revolt is raised. If you're just having preliminary aperitifs – a Dubonnet, a sherry or a sparkling white wine – then the guests will obviously be in a fairly formal mood and it will be difficult to tell which ones are the Communist agitators. So the thing to do is to get some cloth and some bits of old paper, put them down on the floor and shoot everybody. This will deal with the Red Menace on your own doorstep. If you're having canapés, as I showed you last week, or an outdoor barbecue, then the thing to do is to set fire to all houses in the street. This will stir up anti-Communist hatred and your neighbors will be right with you as you organize counter-revolutionary terror. So you see, if you act promptly enough, any Left-wing uprising can be dealt with by the end of the party."

that day would have completely evaporated. From Rome we learned that the unions still "held" Turin, and that serious incidents had not been reported there or in Genoa. Several hours later, this information was directly confirmed to us by the union *leaders*[53] who were there. Fortunately, there had been no deaths among the demonstrators, because that was the piece of good fortune that, deep down, the agitators counted on. In the evening, Milan – the workers' Milan – was discouraged to learn that everywhere else the strike had taken place without incident, but in Rome, and certainly in working class Rome, the events in Milan were perceived in all their seriousness, and they even created more emotion than one could hope for in a capital that is underhandedly insensitive to the impulses of the rest of the country. The city was notified that there was no time to lose, since in Milan neither the unions nor he police had been able to prevent the riot and, even if this riot had, fortunately, been brief, it was only too well known that none of the conditions that caused it had been surmounted, neither in Milan nor anywhere else in Italy. Thus, there was more than good reason to fear that several weeks later, if not sooner, a new riot would turn into a general insurrection.

Instead, three weeks later, on 12 December [1969], bombs exploded at the Piazza Fontana in Milan and in Rome, and in truth we saw the "unforeseeable and traumatic" act of which Nicola Adelfi had written and which so profoundly roiled public opinion in Italy and abroad.

Disoriented and astonished by the number of innocent victims, the workers remained hypnotized by the unexpected event and were led astray by the rumors that followed it – because, confronted by deeds of this type, their spirit is changeable – and, as Tacitus says, "like all multitudes, they were liable to sudden impulses and were now as inclined to pity as they had been extravagant in fury."[54]

As if by magic, struggles that had been so widespread and so prolonged forgot themselves and ceased.

---

[52] English in original.
[53] English in original.
[54] *The Annals,* Book 1, Paragraph 69. Latin in original.

Truthful Report on the Last Chances to Save Capitalism in Italy

Chapter IV:
It is Never Good to Merely Defend Oneself, Because Victory Only Belongs to the Attacker

Before the wars of the French Revolution, this way of seeing things was rather dominant in the sphere of theory. But when these wars, in a single blow, opened up an entirely new world of warlike phenomena . . . one put aside the old models and one concluded that everything was the consequence of new discoveries, great ideas, etc., but also transformed social conditions. Thus one estimated one no longer had any need of that which belonged to the methods of an older time (...) But because, in such reversals of opinion, two parties arise in opposition, the old conceptions find their knights and defenders, who consider recent phenomena to be shocks of brutal force that cause a general decadence in the art of war, and who precisely support the idea that a stalemate – deprived of results, empty – must be the goal (...) This way of seeing things so lacks logical and philosophical basis that one cannot define it other than a pitiful conceptual confusion. But the opposed opinion, according to which everything that happened in the past will not happen again, is also not well considered. A very small number of new phenomena in the field of the art of war must be attributed to new discoveries or new concepts; the majority of these new phenomena should be attributed to new circumstances and social conditions (...) The natural course of war is to begin by defending and to end by attacking.

Carl von Clausewitz, *On War*.

We know that the truth is that much harder to understand the longer it has been suppressed. On the other hand, we have too much experience with the interplay of real forces at the heart of human societies, present and past, to be counted among those who claim, either due to ingenuity or hypocrisy, that one can govern a State without there being secrets or deception. If we thus reject this utopia, we reject no less and just as resolutely the pretention of governing a modern democratic country by founding it exclusively on lies and the systematic use of the *bluff*,[55] as ex-President Nixon, who repented at the end, believed he could do with impunity. Quite the contrary, we have always firmly believed that the people, when they say they want the truth (which the democratic Constitutions give them the right to have), really want nothing other than *explanations*. And why not give them? Why lead them astray in the impasse of the most maladroit lies, as one has done, for example, concerning the bombing of the Piazza Fontana? Our governors, our judges, and those in charge of law and order too easily forget that there is nothing in the world more noxious to power than producing in the mind of the democratic citizen the feeling that he is continually taken to be an imbecile, because this, at bottom, is the spring that unavoidably puts into action the subtle gears of human passions and resentments, by virtue of which even the most timid of petit-bourgeois will rebel and accept and nourish radical ideas. The citizen will then feel he or she is right to demand "justice," less due to a love of justice than the fear of being subjected to injustice in his or her turn.

Today our class politics are in the process of perceiving how costly all the embarrassed and stupid justifications that have accumulated (and always at the wrong moment) on the crucial question of the bombs of 1969 are beginning to be. If there's never been a good politics that has been principally founded on the truth, the worst politics would be exclusively founded on the *unbelievable,* and this because such a politics would incite the citizen to doubt everything, to engage in conjecture, to want to penetrate into all of the State's secrets with a great abundance of casual suppositions and chimerical fantasies. From then on, any imposter would have the keys to the city and could operate with complete freedom and, from the moment that everyone has taken on the figure of shameless artifice, the voter – who habitually contents himself with the plausible – would express with great cries the pretention to know all of the truth about everything, thus hurling a menacing *hic Rhodus, hic salta*[56] at political power. At that point, everyone would be bold

---

[55] English in original.
[56] This Latin expression (a translation of a line in Aesop's fable "The Boastful Athlete") literally means, "Here is Rhodes, jump here." In his preface to *The Philosophy of Right,* Hegel – in an apparent reference to the Rosicrucians – offered an altered translation: *Hier ist die Rose, hier tanze* ("Here is the Rose, dance here"). According to Marx, writing in *The 18th Brumaire of Louis Bonaparte,* "a situation is created which makes all turning back impossible, and the conditions themselves call out: Here is the rose, here dance!"

Truthful Report on the Last Chances to Save Capitalism in Italy

and full of courage in the face of the cowardice with which they would reproach the State, which would be locked into a vicious circle in which it had to successively deny all the preceding official versions of the facts. And it would thus be that a State would inevitably wear out, to the point of losing the strength – we don't want to say the strength to correct its errors, but simply the strength to admit them. Thus, to regain that strength, it would have to expose itself by *finally telling the truth,* because power in Italy is in one of those situations, always dangerous to any State, in which *it is no longer possible to say anything other than the truth.* And the truth, when it finally comes out, after all the lies have been refuted, will be strong enough – although this might also seem unbelievable – to confront all kinds of suspicions and prevail over the general distrust.

> To that truth that has the look of falsehood
> A man should always close his lips, if he can,
> Because he incurs shame where there is no fault:
>
> But I cannot be silence here; and swear,
> Reader, by the verses of this *Comedy* (...)[57]

Goethe was convinced that "writing history is a way of disencumbering us from the past," and we will add that we must immediately and definitively disencumber ourselves from the phantom of the Piazza Fontana, whatever the costs, because the moment has come in which it is infinitely more costly to keep that phantom alive artificially. Moreover, we have wanted this *Report* to be *truthful,* and we wish that the healthy forces in Italy will benefit from the bitter lesson that we must teach ourselves.

Previously [in Chapter III], we detailed the social situation in Italy towards the end of 1969: the workers, without any leaders to obey, were freely acting outside of and *against* democratic legality; they were refusing work and their own union representatives; they did not want (in sum) to renew the tacit social contract on which any State based on rights is founded and especially our republic, which is, according to the first Article of its Constitution, "founded on work." Every day, and everywhere, the workers were effectively violating this Constitution in a hundred different ways. What had been the dramatic choice that our republic found itself confronted with? The choice had been nothing more and nothing less than this: *put*

---

[57] Dante, *Inferno,* XVI, 124-128.

constitutional legality and civil order back into force, or disappear. Who could the State count on to impose the return to law and order at the moment that the forces of Public Safety and the unions were powerless, and the formation of a government with Communist participation was a hypothesis that was rejected as blasphemy by all the other political parties? After the riot of 19 November, the State could no longer count on anything other than its secret security forces and on the effects that their means of information and propaganda could have on public opinion, that is, once public opinion had been sufficiently shaken up by the "unforeseeable and traumatic" bombs of 12 December.

Was the recourse to bombs an error or salvation? It was both at the same time or, rather, the provisional salvation of society's institutions as well as a permanent source of successive errors. This is why we are persuaded that that we can never criticize the operation of 12 December enough, because the bombing of the Piazza Fontana – at the same time that it was intended to be the last warning shot against the menace of proletarian subversion – was already the first cannonball of the civil war, and the manner in which this shot was fired showed the incapacity of our forces in a civil war. The burlesque quality of the successive failed *putsches* of our extreme Right was already contained in that manifestation of great incompetence.

We wouldn't dream of denying the utility for any of the modern countries of similar emergency initiatives, which the necessity of a particular critical moment could impose, just as we would not deny that the bombing of the Piazza Fontana had, in its way, an obviously salutary effect by completely disorienting the workers and the country, and by permitting the Communist Party to rally the workers behind it in the democratic "vigilance" against a ghostly fascist danger, while the unions could finally quickly and efficiently conclude the last and most laborious of the contract negotiations. On the contrary, what we resolutely deny is the idea that these positive effects were assured of or only made foreseeable with a margin of suitable security, that is to say, the idea that we hadn't had recourse to a remedy that was worse and more dangerous than the illness itself when we engaged in an unofficial action in such an inexact way, and this from a double point of view. Above all, too many people were familiar with operations of this type, even before 12 December. Here we will limit ourselves to advancing a single consideration. If just one of the representatives of the Left among all those who knew about it had gone public with the truth that today is on the lips of everyone, even if only as a private person, immediately after the bomb exploded....[58] Well! The television could have said whatever it wanted, but *civil war would have exploded immediately,* and nothing would have been able to prevent it. We can say that it was a real stroke of luck that this didn't happen and at a moment that the political class was surrounded by a sealed but closely watched grouping of murmurs. Moreover, we can reveal that, due to the worst possible choice of guilty parties – someone like Valpreda[59] wasn't believable as the perpetrator of the attack, even if a hundred taxi drivers had, before dying, given a hundred statements for subsequent public display – as well as due to the manner in which the police and the magistrates behaved during the affair, we made this operation into a grotesque farce of misunderstanding and gloom that was more worthy of a South American dictatorship than a European democracy.

Despite all this, how can the operation of 12 December be considered a success? The bombs succeeded in imposing the desired effects to the extent that all of the sources of information put forth, instead of the single true fact, a variety of labels – anarchist or fascist supporters and outcomes – and these sources of information were at first believed, despite or even precisely because of the contradictory versions. On the other hand, the attack also succeeded because one had never seen such reciprocal support by all the institutional forces, such great solidarity between the political parties and the government, between the government and the forces of law and order, and between the forces of law and order and the unions. Thus, what might have appeared to public opinion as an act of parliament "against" the government, the government "against" the bombs, and the bombs "against" the Republic, wasn't obviously a conflict between one constitutional power and another, between the legislative and executive powers, but was well and truly the State itself that, in extreme peril, found itself led to use (as best as it could) certain extreme instruments against itself and for its own support, so as to show everyone that, when the State is in peril, everyone is.

Several years currently separate us from the events that were dangerous to all and sad for some, and that we now criticize publicly. Nevertheless, we must not underestimate what was admirable about this "lyrical expression of history in action" (as Don Raffaele called it) in which the State, reduced to the role of *deus ex machina*, put onstage its own terrorist

---

[58] On 19 December 1969, the Italian section of the Situationist International did precisely that in the form of a wall poster titled *Is the Reichstag Burning?*
[59] Pietro Valpreda, an anarchist who was initially (and falsely) accused of perpetrating the bombing at the Piazza Fontana.

negation to reaffirm its power, because the ruse of reason[60] that governs and moves forward universal history is present in each of its contingent and decisive episodes, even if men do not perceive it immediately, because they are too dominated by the particular passions that serve as pretexts for the permanent conflict that sets them in opposition to each other. Anyone courageous enough to not fear being accused of ingenuity would today be astonished to see how well the expedient of the bombs obtained good effects on the masses, but this hypothetical *naïve person*[61] would be deceived, because, as Machiavelli says, "the majority of men feed upon what appears as much as on what exists; very often they are set in motion more by things as they appear than things as they are."[62] But – and here's the negative limit of similar expedients, also formulated by Machiavelli – "such methods and extraordinary recourses render the Prince himself unfortunate and badly assured, because, to the extent he uses cruelty, his government becomes weak."[63]

Though this might be incomprehensible or terrifying to some people, it is no longer possible to deny the new reality. Beginning in 1969, Italy had a revolutionary "party" that was informal but, consequently, more difficult to strike at. Here, of course, we are not alluding to the para-parliamentary student groups, which wouldn't even frighten the most fearful provincial employee, but all those who, in the factories and the streets, individually or collectively, demonstrated a total refusal of the current organization of work, and even work itself, which in truth was already the [total] refusal of the society that is based upon such an organization. Since 1969, all the acts, failures and successes of our domestic and economic policies are not even comprehensible if one does not put them into relation with the sometimes

---

[60] Cf. Hegel, *The Philosophy of Right*.
[61] French in original.
[62] An alternative translation of the same passage in Chapter XVIII of *The Prince* reads as follows: "Men in general judge more by their eyes than by their hands, because everyone can see, but only a few can feel. Everyone sees how you appear, but few feel what you are."
[63] An alternative translation of the same passage in Chapter XVII of *The Prince* reads as follows: "The Prince must make himself feared in such a way that, if he does not obtain love, he may escape hatred, because being feared and not hated can go together very well, which he will always manage to do when he keeps himself away from the possessions of his citizens and subjects, and their women."

Truthful Report on the Last Chances to Save Capitalism in Italy

open, sometimes hidden conflict that opposes this new reality to all of the traditional institutions, which are now in crisis.

Deprived of leaders, as well as a coherent politics, the workers, young people, women, homosexuals, prisoners, high-school students and mentally ill people unexpectedly decided to want everything that had been prohibited to them, at the same time that they rejected *en bloc* all the goals that our society permitted them to pursue. They refused work, the family, school, morality, the army, the State and even the very idea of any kind of hierarchy.[64] This heterogeneous, violent, uncultivated and clumsy "party" wanted to impose itself everywhere with brutality, and it became, so to speak, *the measure of all things*: that which takes place, since no one can any longer prevent anything from happening; and that which doesn't take place, since our institutions are no longer in a position to make anyone obey them.

To say that this situation has been produced by errors in the management of Italian society would be even more false than unjust – and the Communists know this well – from the moment that such situations can be found in every industrial country, whether they are bourgeois or socialist (as in Poland) – and this the Communists also know well. But such a fact assuredly cannot console us. On the contrary, it is just to say that, in Italy, the *virus* of rebellion found, more than elsewhere, a cultural broth that was particularly propitious for its development, that is to say, a syndrome of pathological infirmities with which our institutions were already chronically afflicted, as we saw in the second chapter of this *Report*.

How have we in Italy reacted to the new revolutionary threat? At first, our politicians simply denied its existence, finding it more convenient to regard the actions of the workers in 1969 in the same manner that they regarded the students of 1968: little more than a phenomenon of morals, a kind of "fashion" that would pass as do all the others. One neglected to consider the fact that a State can temporarily do without universities, which have since then ceased to exist as universities, but it cannot do without factories. Later, when the daily and measurable reality of the damage caused by the social conflict had become striking, our ruling class awoke from its comfortable sleep, believed and judged itself to be besieged by an enemy who was everywhere and that, for this very reason, was difficult to control and define, and from that moment it entrenched itself in a policy of *absolute defense*.

---

[64] Taken from Thesis 12 of "Theses on the SI and Its Time," *The Veritable Split in the International* (1972).

Gianfranco Sanguinetti

In our youth, when we took a course in military strategy, the lieutenant colonel who was in charge gave us a copy of a beautiful book that we still have and that is little known among the men currently in power: Carl von Clausewitz' *On War*. (We should note that the lieutenant colonel's only weakness was being too much of an expert in military questions and too distant from the politics of the regime at the time to have a career in the Italian Army, and the fact is that we haven't heard anything about him since then.) In the 1930s, our own Benedetto Croce[65] deplored the Italian neglect of this work. "It is only the poor and unilateral culture of those who ordinarily study philosophy, their unintelligent specialization, and the provincialism of their social manners that keep them at a distance from books such as the one by Clausewitz, whom they estimate to be foreign or inferior to their discipline." As for us, who, from the moment that this book was offered to us, judged that it was no less important than *The Prince* to a man of power, we would like to quote a passage from it here so as to critique the political strategy of absolute defense that our governments have adopted these past few years.

> What is the fundamental idea of defense? To ward off a blow. What is its characteristic? *To wait for* the blow that one must ward off (...) But an absolute defense would be in complete contradiction with the idea of war, because it rests on the supposition that one of the adversaries commits an act of war; consequently, defense *can only be relative* (...) The defensive form of the conduct of war thus isn't limited to warding off blows, but also includes the skillful use of counter-blows. What is the goal of the defensive? *To conserve.*

And Clausewitz goes on, a little later, to say that,

> The goal of the defensive is negative, it is conservation, while the goal of the attack, *conquest,* is positive, and thus conquest tends to increase the means of warfare, while conservation doesn't (...) The result is that (the defensive) must only be employed to the extent one has need of it, because one is too weak, and that, on the

---

[65] An Italian philosopher, author and politician (1866-1952). His comments on Clausewitz appeared in an essay titled "Succès et Jugement dans le 'Vom Kriege' de Clausewitz," *Revue de Metaphysique et de Morale,* Vol. 42, 1935.

Truthful Report on the Last Chances to Save Capitalism in Italy

contrary, it is fitting to abandon it as soon as one becomes strong enough to be able to attempt a positive goal.

Quite the contrary, to anyone who has observed it with a minimum amount of attention, Italian domestic policy in its entirety, from 1969 until today, appears to be an *absolute defense,* that is, with the sole exception of the use of the counter-attack of 12 December (and we have seen its degree of skillfulness). We would like to specify our thinking here, because it goes to the heart of our critique. All during that year, until its last few months, we had expected (and we could only expect) the aggravation of the crisis. Since the end of June, only the leaders of FIAT – thus proving their foresight – had sought a "global solution" in the negotiations, which nevertheless remained insufficient because one could not hope to resolve a general crisis through an agreement in one sector. What does "to expect" mean? One quickly sees that it means leaving to the workers (who launched the offensive) the time necessary to act in concert, to unite, to reinforce and tighten their ranks. It means letting the unions wear themselves out in a thousand conflicts, during the course of which they were tested daily by the working classes. We do not quite know, and knowing such things now is of little importance, if the roots of the government's excessive wait-and-see attitude were in its conscious *and* erroneous choice or, more precisely, a pure and simple refusal to choose. Nevertheless, we know that this refusal produced almost all of the subsequent errors in political conduct and that, at its basis, there was a crude error of evaluation or, what's worse, a crass ignorance in matters of revolution. In reality, none of the men who were then in government (and who are still there now) believed that it was possible that the workers – without leaders, means or apparent coordination – were capable of constituting a real danger to the security of the State and the very survival of our society order. They simply worried about the economic damages caused by the strikes, which were considered to be enormous, while in fact, in their entirety, they only constituted *the least damage,* because at that moment our economic situation was rosy when compared to the one of today.

On the contrary, we were in one of those circumstances in which the most serious error precisely consisted in not fearing such an adversarial "party" because it had no leaders. One hardly kept this "party" in mind because it was informal and the State was armed, and yet we have always been persuaded (and history only offers us too many examples) that it is fitting to heed populations every time that they take themselves for everything, because *"the misfortune is that their force lies in their imagination and one can truthfully say that, unlike all the other kinds of power, they can do anything they want to do when they have come to a certain point,*"[66] as Cardinal de Retz once said of the Fronde. Moreover, all revolutions in history have begun without leaders and they have ended when they have gotten them.

Thus this absolute defense presupposed that only the workers could carry out "acts of war," to keep to Clausewitz's schema, and this attitude on the part of power gave the workers their principal encouragement. One waited, almost with resignation, and did almost nothing other than wait. Or, more precisely, what one did to justify this attitude led to several laughable episodes of an artificial and useless pseudo-offensive campaign represented by the attacks carried out in April and August. We might admire this monument to political irrationality: these attacks, according to one's calculations or hopes, had won over at least a part of public opinion to the party of law and order at a time when public opinion was generally favorable to the strikers. In that way, one hoped to win the war *with the weapon of public opinion,* joyously forgetting the simple truth that public opinion, when it is hostile to power, harms it, and when public opinion is favorable to power, it does nothing for it as an ally. This was precisely because, at first, one didn't want to understand the nature of the conflict and then because one underestimated the danger, with the result that insurrectional episodes such as 19 November took place. The great fear created by 19 November was thus necessary and sufficient for the change of course in thinking that led to the operation of 12 December, which – having been conducted with such fury – had to be hurried and approximate. In fact, we can say that the time that elapsed between 19 November and 12 December was dominated by the anxiety caused by the approach of an imminent event, which the majority of people imagined would be a riot with much worse consequences than the one in Milan. Every day new authentic or artificial alarms served to put pressure on this or that sector of power or public opinion. A friend whose offices were at Montecitorio[67] reported to us that the entirety of Parliament was so obsessed by the idea of open social conflict, which appeared unavoidable and for which the State was apparently unprepared, that one said that it could read the words *civil war* written on the walls of the auditorium. Following the customs of parliamentary assemblies, what was the most troubling in the depths of their minds was that which one spoke of the least, but they implicitly proved that they didn't forget about it for an instant. Added to this

---

[66] French in original.
[67] The meeting place of the House of Deputies.

Truthful Report on the Last Chances to Save Capitalism in Italy

was the fact that the unshakeable tranquility of the leader of the government[68] was a preoccupation for those who didn't know the reasons for it and regarded it as a kind of unconsciousness. For those who knew the real reason for it, his tranquility was an even greater preoccupation. One knew that the High Commander of our Army, if he was incapable of fighting a classic war, was even more incapable of fighting a civil war and, as for the Army itself, we can say – making use of a recent and welcome expression from a book of "political fiction," written anonymously – "although no one ever speaks about it, our divisions aren't any less disorganized than our postal services."

As we have always found the personality of Admiral Henke to be the least disconcerting, we believed we were authorized at the time to discretely advise him to be prudent and keep himself far above the fray that some politicians had long created around him, so that he would not uselessly compromise either his person or his reputation in the forthcoming chaos, which is always good advice to give to a man so impassioned by action, but so little accustomed to act before having been provided with truly useful and even the most necessary reasons to do so that he always seemed to us ready to undertake noxious and dangerous actions rather than do nothing at all, but this advice wasn't heeded, like all advice that goes against human nature! What followed confirmed this.

It is precisely because one did not foresee the situation in which the operation of 12 December became necessary, and because one then conducted it in such a maladroit fashion, that we have imperceptibly made it a habit here in Italy to confront all the critical situations in the years that followed with the false card of artificial terrorism, which has been lacking believability and especially usefulness. Because the expedient of bombs obtained good results the first time, one has – without posing other questions – made this tactic into a unique strategy, which has since become known under the names "the strategy of tension" or "the strategy of opposed extremisms." Perpetually continuing to defend itself against ghostly enemies – sometimes red, sometimes black, according to the mood of the moment, but always badly constructed – our State has never wanted to confront the problems posed by the *real enemy* of the society that is founded on private property and work, and has wasted its time combating the phantasms that it has created and thus creating an alibi that would clear it of its real desertion. The result of this has been that the State hasn't even obtained the people's support for its hardly believable battles. On the contrary, the following has been the result: para-

---

[68] Mariano Rumor.

Statist emergency practices have become completely ridiculous and, as one says, "burned." Once the game became too obvious, the State was even obligated to put its own secret services chief into prison. No one believed that General Miceli would remain in prison any longer than the time necessary to release him. The insolent hypocrisy with which one has accused him was only a prelude to the hypocrisy with which one released him from detention. Great result! The *Servizio Informazioni Difesa*[69] has become the worst scandal in our country. So we will say this clearly, and once and for all: it is time to end the uncontrollable use of unofficial action, which is brutal, useless and dangerous for law and order, even when it shows itself able to safeguard law and order with the most efficient procedures. More particularly, we would like to ask, What have been the actual fruits and the practical utility of each of the acts of terrorism that followed the one committed on 12 December 1969? What was the usefulness of the pre-electoral attack on the publisher named Feltrinelli,[70] who was an inoffensive Leftist industrialist? What was the usefulness of the elimination of Commissioner Calabresi,[71] when today every citizen knows more about the attacks of those years than he did?

Our secret services' alternation between ineffectiveness and hyper-effectiveness over the course of the last few years reveals a worrisome equivocation: those who can remove it don't want to, and those who want to do so cannot. In this matter, the more one knows about the suspicious maneuvers that take place between the scenes, the less one takes the risk of denouncing them, either because the people who have proof of their existence are personally implicated in this vicious circle or because they fear dying like so many witnesses whom one hasn't wanted to call to testify in the trials of the last few years. Moreover, it is well known that every modern secret service is in a position to greatly abuse its secret character and thus its power, arbitrarily enjoying that which goes well beyond what is necessary for the defense of the general interests of society and forcing silence (by one means or another) upon anyone who advances some well-founded suspicion about the practices that are certainly not above suspicion, but then *"is there any hope for justice when the criminals have the power to condemn their critics?"*[72]

---

[69] Defense Information Service.
[70] Giangiacomo Feltrinelli, murdered on 14 March 1972. The circumstances of his death were made to look like he'd blown himself up trying to dynamite an electrical tower.
[71] Luigi Calabresi, the officer in charge of investigating the attack on the Piazza Fontana, was murdered on 17 May 1972.

Truthful Report on the Last Chances to Save Capitalism in Italy

The paradox resides in the fact that these are not the means by which public order (blanketed by military secrecy) is maintained, *but the means by which it hasn't been maintained,* because everyone has seen how these methods have generally exacerbated the disorder, that is, when they haven't deliberately created it.

In all the States of the world, a secret service receives its orders from the executive power, but the executive power is (fortunately) not administered in all the other States of the world as it is in our country. Thus, isn't it permitted to conclude that the Italian secret service has become the *two-edged sword in the hands of a fool*[73] of which the Latins spoke? By dint of [too many] helping hands and dramatic turns of events, the majority of the population has become *drugged* and habituated to learning about some new carnage at the same time as the recall to Rome of the inquest into the preceding massacre, or the "recusal from office" of a magistrate who came dangerously close to the truth, with the result that one can no longer hope that the healthy forces of this country are capable of obligating the State to make a radical purification by applying pressure from below. Such a purification is urgent, but it must come *from the top,* and our own public intervention marks the beginning of it, at the same time that it shows the necessity of such an intervention: "there where everything is bad, it is a good thing to know the worst."

The magistracy itself, in which men of great value preside, is governed in such a manner that it currently resembles a poor troupe of traveling actors from long ago that, booed in one place, is always hopeful (always in vain) of finally being successful in another town. If this troupe no longer performs the plays that the public in Northern Italy finds obscene or that Rome finds too audacious, it tasks Catanzaro[74] with constituting a Court of Justice that will restage those plays using the same *libretti,* but they are inevitably suspended shortly after the contrasting prologue because the renown of the preceding failures has preceded the show. A humorist from another century said that

---

[72] Slightly modified quotation from Saint-Just. French in original.

[73] The Latin expression employed here, *gladium ancipitem in manu stulti,* seems to include an allusion to "Operation Gladio," which was the Italian code name for the secret NATO plan in which armed groups prepared to either overthrow Communist governments after they'd been formed or before they had seized power.

[74] Not only a politically "neutral" area, but one in which the geography served to aid security procedures.

one of the principal differences between a cat and a lie is that the cat has nine lives.

After doing something stupid, men ordinarily do a hundred other stupid things to hide it. Our State, still dominated by the same men, doesn't behave like a State, but like men: it seeks to limit the damage of one error by making another, more serious one, and it finally arrives at a situation in which it is no longer possible to do anything other than make errors. As one knows, the defense of a bad cause has always been worse than the cause itself, but the defense of a just cause – *and we have the weakness of believing that our world merits being defended* –, when this defense is conducted without dignity and maladroitly, is in every case a crime that obtains effects that are, on all points, the opposite of what was desired.

On the question of "the strategy of tension" and the unofficial services, it is necessary and fitting that, from now on, we be more radical than the Communists themselves, and it pleases us to summarize our thinking on this question with phrases that are not ours.

> It appears to me that we have come to the extreme point of a great danger when there is no other course of action than choosing between enlightening the people and preparing oneself for combat with them (...) If trouble with the plebeians is to be feared, we do not fear popular disgust any less, and we guard against all the steps and proceedings that could excite them. They could lead to greater evils and not exclusive of more serious and more reasonable troubles.

(Thus wrote Francesco-Maria Gianni, former State advisor to the Grand Duke Pierre-Léopold, in a work from 1792 evocatively entitled *The fears that I feel and the disorders that I dread due to the circumstances that the country is currently in.*)

To conclude, we will say that the dramatic turn of events (that decadent theatrical trick) – and its chronic politics in Italy – have sufficiently demonstrated the impotence of the governors, as well as a general desire to change the scene, the plot and the actors. All the serious problems of 1969 are still before us and, if one speaks less of them today, this is only because other, no less serious problems have been added since then, while the men who have not resolved them are still in power and, at the very moment that we are writing, they are in the process of quibbling at length over the collapse, while it is our very Republic that is failing. *Frailty, thy name is Italy!*[75]

Truthful Report on the Last Chances to Save Capitalism in Italy

Chapter V:
What the World Crisis Is
and the Different Forms in which it Manifests Itself

Troy, yet upon his basis, had been down,
And the great Hector's sword had lack'd a master (...)
The specialty of rule hath been neglected:
And, look, how many Grecian tents do stand
Hollow upon this plain, so many hollow factions (...)
When that the general is not like the hive (...)
The unworthiest shows as fairly in the mask (...)
When the planets
In evil mixture to disorder wander,
What plagues and what portents! what mutiny!
What raging of the sea! shaking of earth!
Commotion in the winds! frights, changes, horrors,
Divert and crack, rend and deracinate
The unity and married calm of states
Quite from their fixture! O, when degree is shaked,
Which is the ladder to all high designs,
Then enterprise is sick! (...)
Then every thing includes itself in power,
Power into will, will into appetite;
And appetite, an universal wolf,
So doubly seconded with will and power,
Must make perforce an universal prey,
And last eat up himself (...)
Troy in our weakness stands, not in her strength.

Shakespeare, *Troilus and Cressida*[76]

---

[75] English in original.
[76] Fearing that the result would be a dreadful series of mistranslations, we have *not* provided an English translation of Guy Debord's French, which was in turn a translation of Censor's Italian, which was (we presume) a translation

Gianfranco Sanguinetti

When the present does not regret the past, and when the future does not appear compromised by the precariousness of a present like ours, men and women live their lives in all its richness. To give an evocative example: in the second half of the 18th century, Venetian society could offer itself the luxury of literally forgetting the masterpieces of Vivaldi and Albioni because of the new masterpieces of Mozart and Lorenzo Da Ponte that had come from Vienna.

But in an epoch in which the poverty of a present that is simultaneously anxious and stagnant announces the coming of a troubled and tragic future; in an epoch in which the rediscovery of the masterpieces of the past, quickly pillaged, hardly consoles us; in an epoch in which poverty, and especially cultural poverty, dominates our societies of lost abundance and assaults all of us – individuals and social classes, the leaders and the led, up to the State itself – everyone seems to fidget in a kind of *"absolute anxiety of not being who he really is,"* as Hegel would say. Thus we witness a strange, generalized and universal alienation, by virtue of which no one can any longer play the very role that defines him. The workers no longer want to be workers; the leaders fear to appear to be leaders; the conservatives hide or keep quiet; the bourgeoisie fears being bourgeois – we wish to repeat that, "when all the ranks are disguised, the most unworthy also cut beautiful figures in the masquerade," and then "the unity and peaceful marriage of the classes" evaporates, because there is no longer a "fixed condition" for anyone.

And, in what concerns the Italian bourgeoisie, which was reminded (unsuccessfully) by Giorgio Bocca[77] that "it wasn't born yesterday," and that it was even the first bourgeoisie in history and the inventor of the bank, today we see it believe every word of its adversaries, accord credence to fashionable Marxism and its predictions (instead of having faith in its own history and

---

of Shakespeare's English. Instead, we have provided these lines as they appear in the original text. But our readers should know that the French version contained two lines that were so different from the original English, and yet so relevant to the themes of *The Truthful Report,* that they could only have been intentional: "the unity and married calm of states" was rendered as *l'unité et le paisable mariage des classes* ("the unity and peaceful marriage of the classes"), and "when degree is shaked" was rendered as *quand la hiérarchie est ébranlée* ("when the hierarchy is shaken").

[77] An Italian journalist and essayist (1920-2011) who authored a controversial history of the resistance to fascism during World War II.

culture, which has been forgotten or ignored), and fill its mouth with quibbling about the proletariat and the most adequate means by which the workers should conduct their own struggles to such an extent that, for a part of our bourgeoisie, in the great sunset of capitalism, of which it speaks, *all cows are red.*[78]

This general crisis of identity, in its turn, is only a particular aspect of the current global crisis, but it does not any less merit our attention for that, and while we are on the subject, we would like to argue *a contrario,* for the benefit of the Italian bourgeoisie, by quoting from (and not providing any commentary on) an eloquent passage from a private letter sent to us by a Russian diplomat, whose name we will not divulge, immediately after the invasion of Czechoslovakia in 1968.

> It is stupidity that causes you Italians to raise the question of the workers. I absolutely do not see what you would like to do with the European worker after you have turned him into a question. If you want slaves, you are crazy to grant to the workers that which makes them masters; but you have destroyed, down to their seeds, the instincts that make the workers possible as a class, that is, *that which makes them admit this possibility to themselves.* What would be astonishing if your worker finds that his existence today appears to him as a calamity, to speak the language of morality, as an *injustice*?[79]

---

[78] A détournement of a famous remark in Hegel's preface to *The Phenomenology of Mind* (1807): "To consider any specific fact as it is in the Absolute, consists here in nothing else than saying about it that, while it is now doubtless spoken of as something specific, yet in the Absolute, in the abstract identity A = A, there is no such thing at all, for everything there is all one. To pit this single assertion, that 'in the Absolute all is one,' against the organized whole of determinate and complete knowledge, or of knowledge which at least aims at and demands complete development – to give out its Absolute as the night in which, as we say, all cows are black – that is the very naïveté of emptiness of knowledge."

[79] This alleged letter is actually a détournement of a famous passage in Nietzsche's *Twilight of the Idols*: "I simply cannot see what one proposes to do with the European worker now that one has made a question of him. He is far too well off not to ask for more and more, not to ask more immodestly. In the end, he has numbers on his side. The hope is gone forever that a modest and

Gianfranco Sanguinetti

We have reported this morsel, the italics in which were in the original, not out of a taste for anecdotes, but to show that, in the cold and brutal language that is proper to the Soviet bureaucracy, there can sometimes be more truth, sincerity and realism than in the Marxist dissertations of some more or less intellectual members of the Italian bourgeoisie. All the same, it would be the height of historical irony if our politics, forgetful of people like Machiavelli, must seek its science lessons from the dominant bureaucracy in Moscow! And yet, in Moscow, the power-holding class seems to forget its own identity less than we do ours, and, despite its immense deficiencies, it is aware of its interests, it knows how to defend them, and it knows *against whom* it must defend them. In Russia and elsewhere, the Communists in fact know better than others in the world today that no true revolution is possible if it is not really proletarian, that is to say, if it does not turn against all domination and all ruling classes, and thus against the ruling class that they themselves constitute in the country where they hold power, and it isn't by chance that their political parties abroad have everywhere ceased to speak of a revolution that they cannot in fact accept, because in Russia in 1917 they knew one directly and, if they managed to seize power, it was only by ruining that revolution that the Communists were able to remain at the helm of the State and the economy.

But now, since we are broaching the most important question that we would like to deal with briefly in this chapter, we will say that it has only been since the autumn of 1973 – and here our reference point is the most recent Arab-Israeli war, which was so full of consequences – that the social crisis, which has in the last five years broken out in almost all the European countries, and not just in those countries, has become completely *global and*

---

self-sufficient kind of man, a Chinese type, might here develop as a class: and there would have been reason in that, it would almost have been a necessity. But what was done? Everything to nip in the bud even the preconditions for this: the instincts by virtue of which the worker becomes possible as a class, possible in his own eyes, have been destroyed through and through with the most irresponsible thoughtlessness. The worker was qualified for military service, granted the right to organize and to vote: is it any wonder that the worker today experiences his own existence as distressing — morally speaking, as an injustice? But what is wanted? I ask once more. If one wants an end, one must also want the means: if one wants slaves, then one is a fool if one educates them to be masters."

Truthful Report on the Last Chances to Save Capitalism in Italy

total.

This crisis is global because, *extensively*, all the regimes and all the countries of the world – in one fashion or another – have been struck by it simultaneously, even if the specific characteristics of the crisis had initially presented different predominant threats in accordance with the respective situations of those different countries.

On the other hand, this crisis is total because, *intensively*, it has been the basis of life – insofar as the crisis has unfolded in the interior of each of these countries – that has been subjected to the contagion.

Whether it is a question of political or economic crisis, the chemical pollution of the air that one breathes or the falsification of food, the cancer of social struggles or the urbanistic leprosy that proliferates where there used to be towns and countrysides, the growth of suicide and mental illnesses, what is called the demographic explosion or the threshold crossed by the noxiousness of noise, the public order that is disturbed by dissent and bandits – everywhere one bumps up against the additional impossibility of *going much farther* along the road of degradation of what had been the conquests of the bourgeoisie properly speaking.

We must admit it: not personally, but as the inheritors of these conquests, *we have not known how to think strategically.* Instead, here resembling the little people, rather than a property-owning class, *we have thought and lived from day to day,* systematically hypothesizing the continuation of the present while accumulating insolvable debts for the future, that is to say, every day renouncing a future worthy of our past so as to not renounce a few negligible advantages, which are the deceptive advantages of a fleeting present. As the poet from Vaucluse says,

> Life passes quick, nor will a moment stay,
> And death with hasty journeys still draws near;
> And all the present joins my soul to tear,
> With every past and every future day.[80]

Thus our ruling classes everywhere have today been reduced to discussing nothing other than the *expiration* of their mandate – a mandate that (we too often forget) we do not hold thanks to God or the people, but thanks only to our own abilities in the past –, and even this global discussion is more or less

---

[80] Petrarch, Sonnet IV, in *The Sonnets, Triumphs and Other Poems*, edited by Thomas Campbell (1879).

reduced to the sad examination of the palliatives that would best *delay* this expiration. And this because, in such a process of decadence in action, we have come to the point of total incompatibility insofar as the social, economic and political system that we manage appears to want to tie its fate to the incessant continuation of a growing and already intolerable deterioration of all the conditions of existence for everyone. One has said that the crisis caused by the oil embargo, and then by the increases in the price of crude oil made by the oil-producing Arab countries, has in turn caused the very serious economic crisis upon which the world debates, and there's something true in this observation, but it is only a part of the truth and certainly the most contingent part, even if we cannot say that it is a passing phenomenon. With respect to the current global crisis, we must say what Thucydides said of the Peloponnesian War, "Την μεν γαρ αληθεστατην προφασιν, αφανεστατην δε λογω,"[81] which is really "the truest cause, but the one least spoken about openly," because the real crisis today – which no one speaks about – *is not an economic crisis*, like the one in 1929, for example, which we were capable of overcoming (and we know how). Above all, our crisis is a *crisis of the economy*, which means the economic phenomenon in its entirety, and it is within this general crisis that a particular, oil-related, economic crisis has subsequently appeared.

This is the most worrisome effect of a converging double process: on the one hand, the workers, who have escaped from the framework of the unions, are imposing on us working conditions and incessant salary demands that seriously disrupt our decisions and the forecasts of our economists; and, on the other hand, these same workers as consumers appear completely disgusted by the goods that they willingly purchased not so long ago, thus creating difficulties – if not obstacles – to the circulation of commodities. The result is that we find ourselves in an *impasse*:[82] we are not successful at selling the commodities that the workers refuse to produce or consume. At the root of this crisis, there is not – as some people think – the subjective attitude of the individuals involved, who, nevertheless, are brought into the process and subsequently increase the damages. The economy has entered into crisis *on its own* and, through its own movement, it is misled down the road of its own self-destruction. It is certainly not quantitatively that the economy has

---

[81] Ancient Greek, which Censor himself translates by the phrase that immediately follows it. Cf. *The History of the Peloponnesian War*, Book I, Paragraph 23.

[82] French in original.

Truthful Report on the Last Chances to Save Capitalism in Italy

discovered itself to be incapable of increasing production and developing its productive forces, but *qualitatively*.

The development of this economy, the crisis of which we are the shareholders, has been anarchic and irrational. We have followed archaic models that would be more suitable to an agrarian economy than to an evolved industrial economy, because – like all the ancient societies, which always struggled against actual shortages – we have pursued the maximum degree of purely and progressively quantitative productivity, "not discerning the overflow of what is sufficient."[83] This identification with the agrarian mode of production was then transferred to the pseudo-cyclical model of the superabundant production of commodities[84] in which one has deliberately created "built-in obsolescence" to artificially maintain the seasonal character of consumption, which in turn is used to justify the incessant repetition of productive effort, thus preserving the proximity of shortages. And this is why the cumulative reality of such production, which is indifferent to both utility and wastefulness, today returns to haunt us in the forms of pollution and social struggle,[85] because, on the one hand, we have poisoned the world, and, on the other, we have thereby given to the people – for every instant of their everyday lives – a special reason to revolt against us, who are the ones who have poisoned life. In the last chapter of this work, we will present several remedies for this "economic sickness."

We note here that our power, which from the first symptoms of the new social war has defended (not too well) the abundance attacked by subversion, must today defend *lost abundance*. In sum, we must manage the world's misfortune. Hopefully our readers will be attentive to the following paradoxical coincidence, which is unprecedented in universal history. At the very moment in which the powers of the world are disposed to come to each other's aid – despite their differences concerning *details*, which no longer truly set them against each other – each one of these powers has such great need of help that none of them are in a position to effectively help any of the others. The power of each State is very limited outside of its own borders, because each one is seriously compromised within them.

---

[83] A remark by Francesco Guichardin (1843-1540), an Italian historian and politician.
[84] Cf. Guy Debord, "Time and History," *The Society of the Spectacle* (1967).
[85] Cf. Thesis 17, "Theses on the Situationist International and Its Time," *The Real Split in the International* (1972): "Pollution and the proletariat are today the two concrete aspects of the *critique of political economy*."

On the other hand, the so-called peaceful coexistence between the great powers is not at all the fruit of a commendable choice that was deliberately made in the sphere of global politics, nor was it the result of the successes of modern diplomacy, as the people of the world believe. We know that peaceful coexistence is not a virtue, *but a necessity,* and a much less joyful one than people would like to believe, because if global conflict has no place in these hypotheses, this is not because of the danger that thermonuclear weapons represent, but because of the new and (according to us) more serious social conflict that each nation must attempt to surmount on its own. We can say, in a few words, that global war is no longer possible because peace has abandoned this world and that the highest degree of military power ever attained corresponds to the highest degree of impotence.

Clausewitz said that "war is the continuation of policy by other means," but even this definition, valuable until now, is no longer valuable (and it will not be in the future) because today's alleged "peace" is in fact *the continuation of war by other means,* and it is the continuation *of another type of war* that the States have neither chosen nor declared. Their very weapons must be quickly and completely redesigned following the English example of the professional army, but trained to fight domestically against subversion, while the secret services will henceforth (from a military point of view) have to principally occupy themselves with domestic politics and not politics abroad (but hopefully not following the example of the Italian S.I.D.!). The next "great war" will be a generalized civil war, and it will thus welcome theoreticians capable of instructing professional units that will be engaged in combat "for hearth and home."[86]

Naturally, there will still be wars between the States, but they will be "local wars," such as those fought in the Middle East,[87] and the great powers will have to intervene in them indirectly to limit the damages and counter-attacks on the global level, where these conflicts are liable to involve the advanced industrial countries, which are all in precarious positions. And here it is important to emphasize the failure of the policies of the great powers, and consequently the entire world, after the Arab-Israeli War of 1973. The Israeli victory, applauded by Europe, was obtained with the military and diplomatic support of the United States (as everyone knows), and it cost, and continues to cost, the United States and all of its allies much more than a defeat in the global theatre of operations would have. At that moment, even those who

---

[86] Latin in original.
[87] Cf. "Two Local Wars," *Internationale Situationniste* #11 (October 1967).

were the most reluctant to admit it were convinced of the vulnerability of our entire economic and monetary system, which had already been put into a very delicate situation by the social crisis.

David Ricardo defined wheat as "the only commodity that is necessary, as much for its own production as for the production of every other commodity,"[88] because, in the economy of that time, wheat assured the survival of the laboring forces themselves in a privileged manner. Times have changed, and today it is petroleum that can be defined as *the product that is necessary and indispensible for the production and consumption of all the others.* At the time of the Yom Kippur War, it was enough for Europe to foresee the possibility of spending the winter without heat for the Atlantic Alliance – created to resist the armed forces on the other side of the Iron Curtain – to melt like snow in sunlight. Only Caetano remained loyal to NATO, and today NATO can no longer count on him.[89]

Later on, the energy crisis, the successive increases in the price of crude oil and all the displacements of the economic and financial equilibriums produced – within the crisis of the economy – the current intensification of the economic crisis and, at the same time, we offered to the Arab countries the sword of Damocles that, for our comfort, they have quite willingly been tasked with holding, suspended, over our industries. In passing, we note the mental debility that can be seen in the economic-political calculations of those who have directed our affairs for the last generation. If we wanted to pursue *this* precise form of expansion, which is largely based on low petroleum prices, then we should have maintained the old form of colonialism, and should not have sacrificed it in favor of illusions of immediate profitability from "neo-colonialism." Almost 30 years ago, the troops of the principal bourgeois States controlled almost the totality of the countries that produced our raw materials and sources of energy. Through the most simplistic calculations, we chose to abandon these colonies *at the cheapest possible costs* and we did this *to develop our technology as if we still controlled those countries!* A dozen permanent colonial wars would not have cost us a quarter of the costs of the current predicament.

Moreover, this hardly unforeseeable failure came at the moment when American power over the world had begun to decline, and this failure

---

[88] It was *Karl Marx*, not David Ricardo, who said this, and Marx wasn't speaking of wheat, but of *human labor-power.*
[89] Marcelo Caetano, the Prime Minister of Portugal, was deposed by the revolution of 25 April 1974.

intensified the domestic political crisis, which soon after overthrew Nixon, who departed in ridicule, and it brought beyond the danger level the crisis that for years that had silently torn America's internal social tissue. Thus, the first effects of all these errors were felt right away, but we have only just begun to see them, and we have not seen the end of them. And what can we say about the naïve casualness with which Nixon's successor, Gerald Ford, proclaimed the following in his first speech as president? "Henceforth we know that a State strong enough to give you all that you want is also a State strong enough to take away all that you have."[90] But what do we know? Today, just a few months after this bold declaration, we know that the federal deficit has grown vertiginously since then, and that Ford hopes that, in the budget for the year 1975-1976, the deficit will not exceed 900 percent of the one from the preceding year. If the poor thinkers of a power that grows poorer in the blink of an eye foresee good things, they see badly, and if they foresee bad things, they see quite well. For example, Henry Kissinger, although he is not a "man without qualities," resembles Musil in his defects.[91] He constantly dissolves action in the vanity of action, and the useful in the useless. In other words, like the majority of those with whom he meets every day all over the world, Kissinger lacks a strategic vision of what must be done and what must be avoided – beyond contingent obligations – to save a world that controls itself with a growing difficulty, because it is useless to want to dominate that which has fallen into ruin, when, instead, it is a question of saving that which one wants to dominate. And, concerning the war that the Israelis won over the Arabs, it is enough for us to say to all the modern incarnations of Metternich[92] that they had better reacquaint themselves with a couple of old maxims. First, "it is never a wise course of action to reduce the enemy to despair" (Machiavelli); second, "those who know how to win are much larger in number than those who know how to make good use of their victories" (Polybius).

As for Europe, which seems to have forgotten that it produced all the masterpieces of human thought, and which for the last 30 years has placed

---

[90] Speech given on 12 August 1974. In point of fact, Ford referred to "a government big enough," not "a State strong enough."
[91] Cf. Robert Musil, an Austrian novelist, author of *The Man Without Qualities* (1942).
[92] Clement Wenceslas Lothar von Metternich-Winneburg-Beilstein (1773-1859) was a German-born Austrian diplomat. The Revolution of 1848 forced his resignation.

Truthful Report on the Last Chances to Save Capitalism in Italy

more confidence in the thinkers from across the Atlantic Ocean than it has had in its own: today it is obvious that Europe has fallen apart *even as a simple "economic community."* And, in Italy – if we consider the fact that the greatest efforts to deal with the crisis undertaken by certain centers of economic and political power have only resulted in laughable attempts to return to the old fascist "solution" at the very moment when the last ruins of fascism have reached their foreseeable ends in Portugal and Greece[93] – well, they can go without commentary.

The politicians can deny it as much as they want to, but today their currency of exchange – the lie – is gnawed away by inflation, even more so that the lira: one epoch is over and a new one has begun. We know that men, who are so often ready to interpret the past in new terms, are also frequently brought to interpret the new in old terms, and thus they do not understand what must be done, because changes in the times always and above all express the fact *that the hour has come.* The cohabitation of one epoch with the one that follows it never risks becoming institutionalized in marriage, no matter what is thought by Senator Amintore Fanfani,[94] who would indubitably be more highly esteemed as an interpreter of the Tuscan landscape than history.

But we can say everything that there is to say about the intellectual poverty that is durably installed in power in our country (and that devastates it) when we review the apparently innocent reflections about the expectation of some unknown panacea with which they [try to] amuse us and that abound in our newspapers (and not only in the worst ones). Here, for example, we are thinking of the candor with which our most important daily newspaper has repeatedly stated that it "envies the French for Giscard d'Estaing." It is quite true that our political class, considered as a whole (and with all due exceptions noted), would bring shame to a tribe of Pygmies, but, all the same, this is not sufficient reason to mock our neighbor, unfortunate France, by pretending to envy it for politicians with whom no tribe of Watusis would be contented. Someone who has less urbanity than we do, but whom has had occasion to dine once or twice with the French neo-President, came to conclusions about this person that are not too different from what My Lord

---

[93] Just four months after these lines were written, fascist Spain could be added to this list.
[94] An ex-fascist and Christian Democrat, Fanfani (1908-1999) led an unsuccessful campaign to repeal the laws that allowed married couples to get divorced.

Gianfranco Sanguinetti

Niccolo said in his *post mortem* epigram about the Gonfalonier:

> The night that Piero Soderini died,
> His soul came to the gates of Hell.
> Pluto cried out: 'You, in Hell?
> Foolish soul, go to Purgatory
> With the other children!'[95]

Pardon us for the literary device but, in the current generalization of bad morals, each instance of stupidity asserts the rights that are due it, and imbecility never goes without a patron. Here in Italy, we respect too many [unworthy] things to be worthy of being respected. At bottom, it is not even Giscard whom this journalistic triviality envies the French for having; she envies the enticing *image* of the president-manager, the efficient and hopeful technocrat who casually makes a few spectacular changes in protocol and promotes with juvenile fervor a hundred innovative details that momentarily distract his country from the coming subversion, which in fact still smolders under the ashes, seven years after May [1968].

The "Italian question" – or the French or the English questions, for that matter – certainly cannot be resolved by replacing Flaminio Piccoli[96] or [Mariano] Rumor with someone more "telegenic," less implicated in the failures of the past or less compromised by association with the Mafia, as is Minister Gioia.[97] No one can deny that it is necessary and, at present, urgent to *also* change the majority of the men who must defend our interests, but to replace them with people like Giscard would be a remedy that would not fight the sickness at all. The sickness from which we suffer is spoken about, discussed and written about by the very people who, pretending to be doctors,

---

[95] Machiavelli, *La Mandragola* (1524).

[96] Flaminio Piccoli was the General Secretary and President of Italy's Christian Democratic Party.

[97] In 1973, Giovanni Gioia (1925-1981) was the Minister for Parliamentary Relations. In the 1950s and 1960s, he openly worked to bring members of the Mafia into the Christian Democratic Party. Salvatore Lupo's book *Storia della mafia: dalle origini ai giorni nostri* (1993) quotes Gioia as saying, "*Il partito ha bisogno di gente con cui coalizzarsi, ha bisogno di uomini nuovi, non si possono ostacolare certi tentativi di compromesso*" (The Party, needing new members, needs to unite with people with whom attempts at compromise cannot be prevented).

Truthful Report on the Last Chances to Save Capitalism in Italy

suffer from it: their diagnoses are always diseased and their prescriptions are only additional symptoms of the collective disease. The opinion of Manzoni[98] was that, "we common men are generally made thus: we revolt with indignation and anger against mediocre evils and we are resigned to the extreme ones; we support – not with resignation, but stupidity – the heights of what we had at first declared to be unsupportable."

We will not hide from our readers that speaking so coldly is a thankless task, but speaking otherwise seems impossible and silence would be shameful. And our very coldness in treating the things that touch us so personally is not the product of cynicism, which some malicious minds would like to attribute to us, but the necessity of keeping our cool in the face of the danger that *our* world might be at an end. By contrast, those who do not sense that danger will never be in a position to truly put an end to it.

Those in Italy and elsewhere who currently put forth risky forecasts concerning the economic "recovery," feigning to believe that this crisis resembles unfavorable but fleeting circumstances in the past, do so with demagogical intentions, estimating that it is useful to make the people (to whom they can no longer promise mountains and miracles) believe that *at least the leaders*, unlike the workers, foresee a certain recovery in the next year, but, with each passing fiscal quarter, these same prophets are unavoidably obligated to delay or cancel such unfortunately chimerical changes. The illusion of change then only causes a change of illusions. Piero Ottone[99] recently wrote, and with good reason, that

> the expectation of a misfortune is oppressing and unnerving. When the misfortune finally strikes, we almost sigh with relief and, paradoxically, we suffer less than before. Until yesterday, we feared that the country would collapse; the simple fact that it still hasn't procures a curious sensation of victory for those who were the most pessimistic.

We, who are neither pessimistic nor optimistic, do not even envy those who possess this "curious sensation of victory," but as we do not want to leave too much of its bad mood with the readers who have reached the end of this hardly cheerful chapter, we will provide a little pleasantry, the spirit of which

---

[98] Alessandro Manzoni (1785-1873) was an Italian poet and novelist. He considered the "father" of modern Italian.
[99] Leftist editor of *Corriere della Sera* and correspondent for the BBC.

is not foreign to its subject matter. The pleasantry, which is a very minor Italian art, but the only one that remains alive today, exists in an inverse proportion with the times: the happiest ones come from the most unfortunate days and hold out to them a kind of unique consolation. "It is a shame," the president of one of our most famous national industries said to us, "that pleasantries are not quoted on the Stock Exchange!"

Here's a little story, set in another time and place. The chief of a tribe of Sioux, after a year in which the harvest had been destroyed by catastrophic rainstorms, united his tribe at the beginning of winter to tell them the news. Not knowing how well his anxious audience would take it (they suspected the existence of the calamity), he found an oratorical expedient that our politicians would envy. He said, "My brothers, I have *two* bits of news to announce: one is good, and the other is bad. Let us begin with the bad news. This year you will have nothing to eat but shit. And now the good news: as compensation, there will be enough for everyone."

Truthful Report on the Last Chances to Save Capitalism in Italy

Chapter VI:
What the Communists Really Are and What We Must Do With Them

Princes (...) have found more loyalty and usefulness in the men who, at the beginning of their reign, have been held to be suspect than in those who then had their trust (...) I would say only this, that the prince will always be able to gain with very great ease those men who in the beginning of a principality had been enemies, who are the kind who need to lean upon others to bear themselves up; and they are more powerfully forced to serve him with faith, insomuch as they know it more necessary for them to erase with works that sinister opinion which one had of them. And thus the prince always draws greater utility from them than from those who, serving with too much security, neglect his things.

Machiavelli, [Chapter XX] *The Prince*

At this point of this pseudonymous work, there will certainly be people who, in the course of their reading, have recognized our hand behind a good number of the preceding arguments. We do not want these readers, reading what follows, to change their opinions because, if they have divined from whom this exposé emanates, what comes now is only *apparently* in contradiction with our prior stances and, moreover, was already announced in the preface to this pamphlet. If it is true that, in the past years, if not the last few months, we have said and repeated in answer to the "Communist question" the celebrated phrase by Phaedra's fox, *they are too green*,[100] we must now make it clear that the fox had his reasons to say this, just as today there are other reasons to change one's mind about everything. In truth, it is not at all a question of a subjective change of opinion on our part but, rather, the objective occurrence of the possibility for a useful and necessary change that we – in the company of other, no less qualified people – have been tasked

---

[100] Latin in original.

with preparing, and have been so tasked since the time when it still seemed appropriate for us to emphasize the disadvantages of that change. There is nothing in the world that has its decisive moment, and the capstone of good conduct, especially in politics, is recognizing and seizing the moment.

We will not say that this change of premise creates novelties in the treatment of a question that, in fact, is not new: we will say what is necessary and what has become urgent. For those who have had the occasion to know us in the past, what will be new here will only be our current disposition towards the Communists, which has in fact shown through in the preceding chapters. The hour has come in which it is both necessary and possible to reject a large part of the defects in our nation: the ruse that suits the current situation is doing without one; intelligence consists in never forgetting it; and, in this case, what is prudent is not having too much prudence. At such a moment, it is more important to pay attention to not missing the opportunity, which excellently makes the most of a hundred others in different directions, because "neither the seasons nor time wait for anyone."[101]

Henceforth are finished the seasons of verbal games of prestige in which our political trapeze-artists measure themselves in "parallel convergence" with the Communists, offering them what has been called the "strategy of attention," i.e., the indefinitely long wait before reaching the "historic compromise," which the President of the Council, the Honorable [Aldo] Moro, has defined – with the precautions that have obligated him to walk on eggshells – as "a kind of meeting half-way, something new, which both is and is not a change in the roles of the majority and opposition [parties], the outlining of a diversity that doesn't consist in a change of leadership, but in a modifying addition of the Communist component to the others." *So much noise just to make an omelet!*[102]

Among the political *leaders*[103] who for months have gargled with the "historic compromise" so as to ward it off, no one has yet spoken of the principal and simplest truth in the matter: the "historic compromise" is, in the true sense of the term, *a compromise only for the Communists* and not at all for us. For us, the agreement with the Communists is not at all "historic," at least if we want to call "historic" any *tactical* action that we find it necessary to take to make those who do not want to work go to work. But in this case, and

---

[101] Baltasar Gracian, Paragraph CCLXIX, "Make Use of the Novelty of Your Position," *The Art of Worldly Wisdom* (1637).
[102] French in original.
[103] English in original.

Truthful Report on the Last Chances to Save Capitalism in Italy

lacking such an agreement, how many "historic charges" will our police forces have to lead at the factories? And with what results? Even the former Minister of Labor, the Socialist [Luigi] Bertoldi – who was considered by a right-wing journalist, Domenico Bartoli, to be "a subtle interpreter of the Hegelian dialectic" – said it better than anyone else, and once and for all: "We must decide if we want to govern through the unions or the carabinieri." Because that is the heart of the question, which is as much political as it is economic, because – throughout the last few years – we could have gained much more if we had been able to use the unions three times more than we used the carabinieri. Alberto Ronchey,[104] who is far from the best Italian editorialist, recently wrote that the greatest economic problem we currently face is convincing people to work, and he was right. At present, it is no longer possible to allow ourselves to live by always hoping that the workers will delay their smoldering revolt for "one more moment," or that our industry will regain its breath and vigor, although the anarchy of protest still reigns in our factories. Meanwhile, Italy changes governments, one after another; each of them only lasts for several months; and these are governments that are constantly and uniquely engaged in the titanic enterprise of remaining in power a little longer than what appears possible to them, all the while deflecting all questions, even the least important ones, because they would be enough to make them fall. But who today could better impose on the country a period of convalescence, during which the workers would cease struggling and go back to work, than the Communists? Who would be a better Minister of the Interior than Giorgio Amendola[105] when it comes to the eradication of the delinquency that has spread to all levels of society or the silencing of the agitators, either through good or less good methods? We must undertake a long-term governmental effort and, to do this, we must have a solid and resolute government. Today, not accepting a "compromise" such as the one in question in truth means, for us, accepting the fatal compromise of the very existence of tomorrow. We must remember that neutrality in such an affair is the daughter of irresolution and that "irresolute princes most often follow the neutral route in order to avoid present perils, and most often end in ruin."[106] So as to not see the real danger, we feign to believe that the agreement with

---

[104] Employed by *Corriere della Sera*.
[105] Giorgio Amendola (1907-1980) was a deputy in the Italian Communist Party from 1948 to his death. He advocated non-Marxist positions and the making of alliances with other political parties, especially the Socialists.
[106] Machiavelli, Chapter XXI, *The Prince*.

the Italian Communist Party [ICP] is a danger, and we flee before both of them.

Even if they are obliged to admit the justness and utility of what we are saying, timorous spirits may find in our remarks the slight fault that they appear to set little value on the dangerous aspect of placing a Communist party at the heart of political power when, at this stage of the crisis, our powers are incapable of continuing to make the workers work. *Who will guard our guardians?*[107]

We would respond to such an objection that it is without foundation and that fear is a bad advisor. First of all, we must never fear a future and hypothetical danger at a moment when we are dying from a present and certain one, and, moreover, we must never risk all of our fortunes without risking all of our forces. Since the current strength of the Communist Party and the unions has already served us well and, in fact, has been our principal support since the autumn of 1969, and since the effects of this support have, until now, been quite insufficient to reverse the process, there is no doubt that our interests lie in *galvanizing* this strength as a matter of great urgency, and to do so by offering it access to the central point of application in society, that is to say, by introducing that strength into the center of State power.

On the other hand, we will say that the alleged future dangers of Communist participation in our government only exist in the sphere of illusions about the revolutionary tendency that the Communist Party constitutes in our society. These illusions were artificially spread at an epoch that is now over, that is, when they were useful for the defense of a world that (the times having changed) today wants to be defended with the support of these same Communists. Only our current crop of politicians – who, despite their unfortunate failures, aspire to become [permanently] autonomous in their existence as simple delegates of Italian society in the service of its governmental administration – still pretends to hold as a real [and permanent] fact in strategic reasoning that which (i.e., the allegedly revolutionary tendencies of the ICP) was never anything more than an ideological "export" for consumption by the people. Which makes these worn-out leaders subject to this severe condemnation: what they in fact want when they hang on to their old specializations (when necessary modernization imposes "recycling" on them) is not to prolong (for their own limited interests) the apparent existence of a trade that they still know how to ply, but *a trade that they do not know how to ply.*

The Trojan Horse is only to be feared when there are Achaeans inside it.

---

[107] Latin in original.

Truthful Report on the Last Chances to Save Capitalism in Italy

The Communist Party has wanted to manufacture, and even must still try to manufacture, a certain costume to disguise itself as the enemy of our City-State, but *it is not* such an enemy, just as our leader is not Ulysses. In fact, the Italian Communist resembles the carpenter in *A Midsummer's Night's Dream* who lets half of his face be seen through the mask he is wearing and who says to the members of the audience: "I (...) entreat you, – not to fear, not to tremble: my life for yours. If you think I come hither as a lion, it were the pity of my life: no, I am no such thing."

And precisely because we dare to admit that the Italian workers, who have taken the offensive in the social war, are our enemies, we know that the Communist Party is our support. We can no longer continue to reassure the country by pretending that the opposite is true, because we have come to the moment of truth, when lies no longer work and only force will do.[108] In past years, when we happened to speak of the Communists with Raffaele Mattioli, we never heard that he found them worrisome, and many times we heard him repeat the same conclusion: "They are quite brave." When Togliatti, a year before his death,[109] sent his last book to Mattioli, he (both flattered and amused) showed to us the dedication, which was written in the famous turquoise-blue ink of the Communist *leader*[110] whom imbeciles feared and we appreciated: "To My Friend (...) with the only regret that I cannot call you Comrade," if our memory serves us well. Who knows if Raffaele Mattioli, were he still with us, would not, in his turn, have written a dedication of the following type: "To Comrade Amendola, in the hopes of soon being able to call you 'Your Excellence.'"

In any event, we will not let ourselves forget that, for a long time, our parliamentary majority has ruled with the Communist opposition, and that the Communist opposition has been opposed to the same things to which the majority has been opposed. And yet today the entire political life of this country is paralyzed by the simple idea – a nightmare to the Christian Democrats – of granting a few administrative posts to the Communists. Until quite recently, the Christian Democrats found semi-rational justifications for the necessity of their keeping a monopoly on power by continuing to hide the manner in which that power has been managed, as well as by hiding several particular facts that were so scandalous that, if they were known, would have immediately caused the immediate dissolution of their party. But now that

---

[108] Machiavelli, Chapter VI, *The Prince*.
[109] Palmiro Togliatti, a leader of the Italian Communist Party, died in 1964.
[110] English in original.

these facts are, little by little, becoming known throughout the country, these justifications have become null and void, and it is the dissolution of Italy itself that we must avoid, if we can.

Let us pose the question: *What is the alternative* to the "historic compromise"? Sooner or later, we will be in a situation in which neither the Communists, the unions, the forces of law and order, nor the secret services will be able to prevent the workers from mounting a general insurrection, all the consequences of which are difficult to foresee. In the best of hypothetical situations – and we only see two of them – if this insurrection does not become a pure and simple civil war, that is to say, if the Communists succeed in taking command of it (first by seeming to participate in the insurrection and then by seizing command of it), it is obvious that Berlinguer[111] would be able to set his conditions, and he would not be disposed to sharing his government with us. Riding the crest of the insurrectionary movement, the Communists would seize control of the State in the name of the workers, whom they would call upon to defend it. But, on the contrary, what seems to us more probable is that the credibility of the Communist Party among the workers would be completely exhausted at the moment of an insurrection – this is quite foreseeable – with the result that the Communists' attempts at "recuperation" among the ranks of the insurgents would be useless or impossible. Civil war would no longer be avoidable, and the Communist Party, amputated from its base, inevitably made up of revolutionaries, would no longer be of any use to us. These are the two variations that form the single alternative to the "historic compromise." *There is no third one.*[112]

During such an event, what would become of the Atlantic Alliance, which is already in a state of crisis? And what about the Warsaw Pact, which was powerless during the workers' insurrections in Szczecin and Gdansk?[113] In the tragedy that would follow and play itself out in a theatre of war that would be no less vast than the territories affected by the current crisis, we would only be able to repeat – in the guise of a useless *mea culpa* – this verse from Aeschylus' *Agamemnon*:

Where, where does the Law hide?

---

[111] Enrico Berlinguer (1922-1984) was a leader of the Italian Communist Party. He favored a "Euro-Communism" that would be separate from the Soviet Bloc.
[112] Latin in original.
[113] December 1970.

Truthful Report on the Last Chances to Save Capitalism in Italy

Reason despairs of its powers,
Intelligence gropes numbly,
Its swift resources are dead.
Our rule is compromised,
Disaster is near:
Where can I turn?[114]

In sum, our opinion today on the "Communist question" can be summarized in a single phrase: We do not make a question of that which is no longer one, while the real questions and problems do not wait upon the decisions of Senator Fanfani, who is *slow in providing what may prove of use,*[115] to get irremediably worse. Giovanni Agnelli[116] – who is, among our young men of power, perhaps the only one who can flatter himself with possessing an intelligence that is the most deeply rooted in the reality of our epoch – has openly offered the same analysis that we have put forward. Despite certain differences in the details, our views converge where the majority of the conclusions are concerned. Without saying anything about our private commitments, we will content ourselves with recalling to our readers one of his publicly stated positions, enunciated at the beginning of 1975:

> If our sickness is nearly fatal, we are allowed to think that the Communist Party has understood the necessity of making good use of it, so that we can all save ourselves together. So that class hatred does not come to set the world on fire and divide it into two parties: the enragés in the streets and the others in their *bunkers*[117] with their bodyguards.

We could not say it any better ourselves.

Finally, let us conclude. With the aid of the Communist Party, we will either succeed in saving our domination, or we will not succeed at all. If we do succeed, we will easily dismiss the Communists, as well as a large part of the current political personnel, as if they were domestic servants. The Communists themselves have already clearly accepted this as an article in their work contract, and we have known since Heraclitus that "all that crawls

---

[114] In point of fact, this verse is *not* from Aeschylus' *Agamemnon*.
[115] A line from Horace's *Art of Poetry*. Latin in original.
[116] The director of FIAT Motors (1921-2003).
[117] English in original.

upon the earth is governed by blows." And if we do not succeed, nothing else will matter, because everyone will admit that it would be the worst of byzantine discussions – when the Turks are at the ramparts – to calculate which trophies are going to be awarded to the Greens and the Blues at the circus,[118] in a world that will have collapsed.

---

[118] Factions in the Byzantine chariot races, circa the Fifth Century CE.

Truthful Report on the Last Chances to Save Capitalism in Italy

Chapter VII:
Exhortation to Rescue Capitalism from its Irrationalities and to Save It[119]

They find me difficult?
I know it well:
I obligate them to think

Alfieri, *Epigrams*

He who considers the world in accordance with reason is himself considered in accordance with it. We must act in accordance with the times, and they have changed. To want to go against them is an undertaking whose success is as impossible as its failure is quite assured. The proximity of the fateful moment, *if it is eventually perceived as such by us all,* can paradoxically be our last chance for salvation and perhaps one day we can say, in our turn, what the Prince de Condé said during the religious wars:[120] "We would perish if we were not so close to perishing."[121] On the condition that we know how to exploit for our exclusive advantages *all* the occasions that are presented to us, none of the evil will harm us, despite the undeniable precariousness of our current situation. In the words of the "Exhortation to Retake Italy"[122]:

> At present, to know the virtue of an Italian spirit, it has been necessary that Italy reduce herself to the conditions in which she is at present (...) without chief, without order, beaten, despoiled, torn, overrun, and having borne every sort of ruin.

We will say to those who would accuse us of speaking too much or too quickly of our ruin and its non-hypothetical proximity that such is the primary task of

---
[119] Cf. Machiavelli, Chapter XXVI, *The Prince*.
[120] Louis de Bourdon (1530-1569). The French religious wars lasted from 1562 to 1629.
[121] French in original.
[122] Machiavelli, Chapter XXVI, *The Prince.* Latin in original.

those who truly want to avoid it, because one does not always find oneself in the position to avoid such disasters. And, moreover, what else is there to speak about today?

The intelligent conservative can express the principle of his actions in a single phrase: *everything that does not merit being destroyed merits being saved* – and this immediately and everywhere in the world. But that which does not merit beings saved, that is to say, that which is in contradiction with our own salvation or, more simply, anything that is an inconvenience or an embarrassment, must be abandoned and destroyed without beating around the bush or superfluous scruples. Unburdening oneself of the dead weight of the past is necessary to make the task of cleaning up the present less difficult.

Today, the *principal* irrationality of capitalism is that, although it is under dangerous attack, it does not do everything necessary to defend itself. But we will admit that there are others. We must correct them as well, if we can. In those areas in which our management has been unreasonable, it must be changed, because, ever since the origin of the bourgeoisie, all of our power has been intimately linked to *rational management,* and it cannot last without it. There is nothing new about the appropriateness of making profound reforms. We have given birth to them in every epoch. That is our strength: we are the first society in history that has known how to correct itself continuously. We call "unreasonable" everything that is not a real necessity for our possession of society and that produces results that are objectively in contradiction with those necessities, that is to say, results that we ourselves can measure and are felt by everyone. We will mention the necessary reforms below.

For the moment, we must repeat that, in the midst of the current dangers, we must (as the French say) *make every piece of wood into an arrow*,[123] starting with the most accessible and malleable pieces. Thus, we must employ our own Communists – rather than sell the entire country to the Arabs, as some of our insane politicians have seriously proposed – with the sole goal of making the most of this experiment with a government in which the Communists participate. But this experiment will cost us nothing, while the logic of the other proposal would inevitably lead to our complete dispossession. How is it possible to compare, even for a moment, two obviously unequal solutions? What is inconceivable on the plane of logic properly speaking obeys a particular logic that is hidden but easily discernible. Should we be able to save ourselves, three-quarters of our

---

[123] French in original.

Truthful Report on the Last Chances to Save Capitalism in Italy

political personnel must be discharged. Should we fail to save ourselves, these same people will remain in place and, in a few years, they will squander or embezzle a large part of our capital, which they will eventually expropriate from us and without even assuring the power of the new property owners. In the aftermath of this grotesque prospect – which in fact supposes that the productive forces and the properties of Europe would in large part belong to a few Arab potentates, who would control the defective international monetary system because they would provisionally control the principal source of energy upon which the industrialized countries are dependent – would not the workers, from whom we already have so much trouble, expropriate these new foreign, archaic and perfectly incompetent masters with an even greater facility than they would have with us? Transporting the property-owning class of our country to exotic and backwards locations means selling our birthright for a plate of lentils. But could such *upstarts*[124] hope to control our country? With their own troops or with the help of ours? With our political skill or theirs? Our troops are no longer reliable, and theirs are worth nothing. Our skill is worn out. As for theirs: simply posing the question is to answer it [in the negative].

Thus, we will not be surprised if those responsible for such a strategy, especially in Italy, have no other policy than the complete *liquidation* of our national patrimony and its clandestine export to their Swiss bank accounts. While the high functionaries of our government ministries and economic organizations will charge very dearly – in bad money, alas! – to depart from careers that have already departed from them, the hospital in Padua has announced that it will sell to the highest bidder a Mantegna[125] that belongs to it. All of those who are responsible for the management of Italian society, seeing that society march so quickly to its forfeiture, dream of selling what he or she holds. And, in the final analysis, what they hold is Italy itself, its monuments and its soil. And they want to sell it all because soon our productive forces, with such bad workers and such bad managers, will not be worth much on the market. We must counter those who plan to offer Italian society up to a "Public Takeover Bid."

We would like to return for a moment to one of our preceding statements, according to which we must (without scruples) remove all the *impediments* to the surmounting of the crisis in which our State is in. For example, a year ago, President Leone,[126] who is not completely unappreciative

---

[124] French in original.
[125] Italian painter (1431-1503).

of our arguments, made an allusion (with perhaps too much circumspection and, thus, without any success) to the necessity of a constitutional reform that certain Communists now believe to be urgent. Today, we must propose a reform that is both radical and favorable to the restructuring of the Republic in conformity with the highest-priority necessities for the survival of our world and that, of course, would not be prejudicial to the continuation of democracy, which we said was important to us in the first chapter of this *Report*.

With the commitment of the Communist Party, as much in the elaboration as in the application of the new constitution, we are persuaded that there is a real possibility of surmounting this great crisis. The new *Magna Carta* must maintain democracy, yes, but in a disabused way, thus contrary to what has happened in the first 30 years of our Republic. Maintaining democracy means maintaining the rule of the vote, which is the basis of all the free, modern republics. We know that this rule is the inverse of the one that presided over primitive democracy. Among the ancient Greeks, the rule was to count the votes of those who were ready to fight openly for one camp or the other, and Plato (and subsequent history) showed how this primitive democracy led to disorder and despotism. In its modern meaning, "democracy" must, on the contrary, be understood to be the manner of making the people vote on all the questions for which they are not disposed to fight. This aspect must be accentuated, and we must summon the citizens to vote, as in the past, but on a much greater variety of subjects that are not detrimental to the smooth functioning of society, and the citizens must continue to choose between diverse candidates. But these candidates, no matter what side they come from, must have already been selected in their turn, and with a qualitative rigor unknown in our times, by a veritable *elite*[127] in the spheres of political power, the economy and culture.

And this economy itself – this modern technology that we make use of, and whose power is virtually unlimited – requires that we make a better and *more intelligent* use of it. That is to say, we must no longer allow ourselves to dominate through this power, which incessantly tends to become autonomous by escaping from our hands, which in the recent past have manipulated it, above all, according to democratic and demagogical fictions upon which

---

[126] Giovanni Leone (1908-2001), a right-wing member of the Christian Democratic Party, was the President of Italy from December 1971 to June 1978.

[127] French in original.

Truthful Report on the Last Chances to Save Capitalism in Italy

(during the epoch of "the abundance of well-being" and market abundance) we built a giant with clay feet. But since that epoch is over, we must now cease to make the people consume images that are too beautiful and too wild, and must instead give ourselves the means to make them consume images of a reality that is less harsh than the current one: less pollution; fewer automobiles; better bread, meat and houses; and so forth. In sum, the reform of our economy *from the ground up* and its reconstruction on more solid bases must establish a new economy, one that is capable of being *both* authentically liberal and severely controlled by the State – certainly not *this particular State*, because it must be rigorously lead by an *elite*[128] that is really worthy of the name. We will return to this subject below.

Today, it is important for us to consider that we must not only maintain a dominant class, but *the best possible dominant class.* Our government ministers must strive to rule through merit and talent, because we know that those who start out aiming to be satisfied with a secondary position will never attain it: they will never attain anything at all. If today this minimum requirement seems too utopian or too ambitious, it is so with respect to the pitiful panorama of our current crop of politicians. But such a requirement, which the current situation makes obligatory, is not in fact disproportionate to the reality that we must eventually confront and to the long-term tasks that the good administration of our society requires.

What is convenient to a prince that he might be esteemed?[129] Which men are able to save our society? This is what we must ask when we are choosing our governmental ministers; this is what is especially neglected when we privilege a hundred laughable "titles of merit," such as the fact that the Honorable [Aldo] Moro is more or less the enemy of Cefis,[130] or that someone else's wife is the intimate friend of General Miceli's wife. "Stranger," Plato says, "the moment has come to be serious,"[131] and we know the interest that this philosopher had in the political problems of our peninsula.

Well! We will say, and we will try to prove, that today in Italy the men we need *exist,* and we must make use of them as soon as possible, by bringing

---

[128] French in original.
[129] Machiavelli, Chapter XXI, *The Prince*. Latin in original. (In the translation provided by Guy Debord, this phrase is rendered as "How should the prince govern to acquire esteem?")
[130] Eugenio Cefis, the chairman of ENI (petrochemicals) and Montedison (chemicals), both State-owned enterprises.
[131] *The Republic.*

them out of the limbo to which a herd of Christian Democratic notables, disguised as wolves, flatter themselves with having condemned these men forever, so that these same Christian Democrats can have the pleasure of satisfying their own raging hunger for ministerial posts and clients in complete freedom. Moreover, a few traits would suffice to define these men, because merit accounts for so little in our Republic, and a few well-chosen ministers would suffice to make any State function as it should. It is true that in France under Louis XIII, a single one sufficed. But it is also quite obvious that if we want to continue to coat the various pâtés of our governments in Italian-style sauce – by assigning a ministerial post to a man of Bruno Visenti's talents,[132] and another one to someone like Gioia, of whom *it is well to say nothing*[133] – we will compromise to the very roots any possibility of action by men of value, and we will once again prove right Mussolini's justifying formula, according to which "governing Italy is not a difficult business; it is a useless one."[134] Fortunately, the future of capitalism is not tied to the future of Christian Democracy, no more than it was to the future of fascism, but let us recall that a half-century of stupidity in power is a hardly enviable world record, and especially if no one tries to contest it. Because today few and far between are the men of talent who will take the risk of compromising themselves in the midst of the administrative corruption of a State that appears to be, in the words of Dante, "the sad sack that covers with shit everything that it swallows."[135]

To save ourselves from the threat of subversion, which will probably persist in the years to come, even if the Communists in government are able to master it better than we are at the moment, our first operation must not be an obstinate and obtuse defense of current Italy and its incapable leaders. On the contrary, our first operation should resemble a *scorched-earth policy*, which will permit us to unburden ourselves of these men and the frilly trimmings with which we cover our poor Republic. And, simultaneous with this radical housecleaning, we must reconstruct around ourselves a society provided with all the qualities that would render it worthy of being defended in the eyes of many people. And who knows if, at that moment, the workers themselves will

---

[132] Bruno Visenti (1914-1995) was an industrialist who became the Minister of Finance in 1974.
[133] Dante, *Inferno*, IV, 104.
[134] In point of fact, this quip wasn't made by Mussolini, but by Giovanni Giolitti (1842-1928), who'd served as Prime Minister several times in his life.
[135] *Inferno*, XXVII, 26-27.

not cease to attack us so violently, even if they must always remain irreducibly hostile to private property at the bottom of their hearts? But without venturing into utopian philosophical theories about the future of the world in a time when, personally, we will no longer be around, it is fitting to consider, while we are still here, all that would be necessary *to have our world die out.* In the final analysis, who are our real enemies?

We will say that, today, we must confront *several* hostile realities, only one of which is historically immanent to our mode of domination and production: the proletariat, which has a natural and perpetual tendency to revolt. The ancient Romans summarized this fact in the adage *we have as many enemies as there are slaves.*[136] Once we have taken action upon this incontestable and constant fact, it will be important to see if the other realities that are hostile to us have the same immutability and constancy. Even more precisely, we would like to say that it will be fitting to see if these other realities are as necessary and *useful* as the proletariat. Because we should not forget for an instant that the workers, at least when they work and do not revolt, are the most useful reality in the world and merit our respect, for in a certain way they (under our well-informed direction) produce our wealth, i.e., our power. Well! We would contest the idea that the other realities that currently contest our power are in fact necessary and unavoidable. And we propose to examine at least two of them here: the bad morals and incompetence of which our political class have given ample proof, on the one hand, and economic anarchy, on the other. These two phenomena are deleterious, but both can be opportunely eliminated, because they depend on our will.

For those who regard what we define as the "insufficiency" (to speak euphemistically) of our governing strata as a whole, and setting aside all due exceptions, we can affirm that we must no longer have scruples about letting it sink like a stone in the *great sea*[137] of its errors and scandals, because we already have shown it more gratitude than it deserves for the services that we admit that it has rendered us in the already-distant past, and for too long we have accorded it patience at costs that we did not believe that we were capable of sustaining. Because patience, among all the human virtues, is, according to us, the only one that ceases to be a virtue when one practices it excessively. We leave to the Pope, who is less pressed than we are by the contingent necessities of mundane life in this century, the occasion to make an

---

[136] Latin in original.
[137] Latin in original.

act of charity by rescuing and clearing the consciences of these *orphans of power*. Apart from the satisfaction that we must finally provide to public opinion, which is legitimately tired of seeing incompetence in power being rewarded, we can spare ourselves the future burden of having to defend the men who, instead of conducting a policy of intelligent conservatism, as we have required of them, have instead preferred a policy of obtuse reaction that always squanders everything that passes through their hands. These are men supported by our capital, which they have declared that they want to defend so as to mock the voters, and now they support themselves upon the voters so as to mock us. Finally, these are men who (to once again express ourselves by quoting Machiavelli), "while you use them, you lose the power to do so."[138]

Moreover, even in the Christian Democratic Party there are intelligent men, and here we do not simply allude to people like Andreotti and Donat-Cattin. But in good conscience, how can we say that the intelligence of these politicians can bring forth fruit when Fanfani asks them to make use of it with the sole aim of defending the indefensible and the useless, meanwhile systematically neglecting to save the essential? The survival of a political world of this type is already *in itself* one of the hostile realities that we must cease to keep alive. We must *rid ourselves of it, "and the combat [thereafter] will be short."*[139]

As for what we have called "economic anarchy," we will say that, from now on, we must authoritatively limit the tendency to accumulate excessive profits in certain basic sectors where the level of development reached by modern techniques – especially chemical ones – permits everything, but where the results assault the population in its everyday existence and tend to deprive it even more of the little that we must absolutely let it have. For example, we completely disapprove of the industrialists who take the risk of uninterruptedly provoking the people, who are made to consume petroleum-based products, chemically treated wines and inedible food with the sole aim of increasing their sector-based profits, insolently neglecting the more general and superior interests of our class as a whole.

We repeat that nothing more provokes the democratic citizen than the impression that we give him when, with impunity and systematically, we take him for a ride. Even when this citizen is disinterested in politics, he is not

---

[138] Machiavelli, Chapter XVI, *The Prince*.
[139] Petrarch, quoted at the very end of *The Prince*: "Virtue against furor / will take up arms; and the fighting will be short; / for the ancient valor / in Italian hearts is not yet dead."

Truthful Report on the Last Chances to Save Capitalism in Italy

insensible to the quality of what he eats or the air that he breathes. On the contrary, we must preoccupy ourselves with maintaining the best possible qualitative levels of life, primarily for the dominant class and secondarily for the dominated classes. Moreover, in 1969, an industrialist like Henry Ford said (and we would like to quote his own words), "the terms of the contract between industry and society have changed (…) We are called upon to contribute to the quality of life much more than the quantity of goods."[140] Playing the hypocrite does not result in anything good or, at least, it *must no longer* be profitable. We are hardly brought to greet the news of concerning Montedison's balance sheets with the satisfaction that is felt by the poor money-saver who is also a small stockholder, especially when those assets have been more or less acquired by the means that Scalfari has recently revealed to the public in his book *The Master Race*[141] and when these very profits, in truth, represent a formidable incitement to social revolt.

And since we have cited Eugenio Scalfari, a man whose courage and intelligence we value, we will seize this occasion to express our opinion on what he has excellently defined as the "State bourgeoisie."

(Precisely one of the reasons that led us to choose for this *Report* the old form of expression of the pamphlet, instead of a more systematic text, is that we need not reject the pleasure of talking about this and that, as one does in conversation, which allows us to touch upon everything without ever have the pretention of being exhaustive and, at the same, allows us to avoid getting bogged down in the marshes of sophisticated "demonstrations" of which our politicians are fond when they try to pass off their elastic "truths." To say *the* truth, few words suffice: *the truth is the indicator of both itself and the false.*[142] And because this fashion of writing is rapid, it appears useful to us, at a moment when so many other commitments that cannot be put off impose on us the necessity of not wasting time.)

This "State bourgeoisie," which combines the faults of the parasitical and decadent bourgeoisie and those of the bureaucratic class that holds power in the socialist countries, is one of the several products of the "Italian style" of management of power, and it is a highly noxious residue of the parceling out of this power. Cefis, the President of Montedison, is the model that inspired Scalfari's description. But, in reality, this "State bourgeoisie" exceeds this model; it is nested almost everywhere in the nationalized

---

[140] Henry Ford, speech to the Harvard Business School, 1969.
[141] *Razza Padrona: Storia della Borghesia di Stato* (1974).
[142] Spinoza, *Ethics*, I, proposition 36. Latin in original.

industries and those that involve governmental participation, as well as in the forest of the 60,000 public "organizations" in existence today, and thus it possesses a proper power that is autonomous with respect to the large, traditional bourgeoisie, and it has founded on this power what Alberto Ronchey has pertinently called "Christian-Democratic State capitalism." The members of such a "master race" are, in reality, individuals who have no original personal patrimony and no culture. They aren't simply deprived of a culture worthy of a ruling class, but they are, even from a distance, obviously deprived of the culture of an austere petit-bourgeois (a teacher, for example) in the past. Of course, today, only a relatively limited number of these individuals hold real power, and the largest number of them can only do harm with their limited talents. But this does not change the fact that this phenomenon is growing and thus merits our attention.

Over the course of its history, capitalism has continuously modified the composition of the social classes and has done so to such an extent that it has transformed society. It has weakened or recomposed, eliminated or even created the classes that have had subordinate but necessary functions in the production, distribution and consumption of commodities. Only the bourgeoisie and the proletariat have remained the historical classes that have – in a conflict that has essentially remained the same for the last century – continued to play out the destiny of the world. But the circumstances, scenarios, walk-on performers and even the spirits of the principal protagonists have changed with the times.

Thus, this phenomenon is not particular to Italian society. The expansion of the last 30 years, which is unprecedented in the history of the global economy, has involved the necessity of creating a class of *managers*,[143] that is to say, technicians capable of directing the industrial production and circulation of commodities. The *managers*,[144] as one has called them since their modern popularization, these *executives*[145] have necessarily been recruited from outside of our class, which can no longer assume the totality of managerial tasks on its own. Despite a gilded legend, which they are the only ones to believe, these executives are nothing other than a metamorphosis of the urban petit-bourgeoisie, previously constituted in the main by independent producers of the artisan type, who today have *become salaried,*

---

[143] English in original.
[144] English in original.
[145] See Thesis 36, "The Situationist International and Its Times," *The Real Split in the International* (1972).

no more or less so than the workers properly speaking, and this despite the fact that sometimes these executives hope to resemble members of the liberal professions. Given this "resemblance," which they have obtained on the cheap, these executives have in a certain way become the object of the promotional reveries of many strata of poor employees, but, in reality, they have nothing that could define them as rich. They are only paid enough to consume a little more than the others, but the commodities they consume are always the same ones consumed by everyone else.

Unlike the bourgeois, the worker, the serf and the feudal landowner, the executive never feels *at home*. Always uncertain and always disappointed, he continually aspires to be more than he is or will ever be. He pretends and, at the same time, he doubts. He is the man of uneasiness, so little sure of himself and his destiny – not without reason – that he must continually hide the reality of his existence. He is dependent in an absolute manner, and much more so than the worker, because he follows all the fashions, including ideological fashions. It is for him that our "avant-garde" writers and authors make the repugnant *best-sellers*[146] that turn bookstores into supermarkets. We refuse to set foot into such places. (Fortunately there are still several good stores devoted to old books, and these are our consolation.) It is for the executives that, today, one changes the physiognomy and functions of our towns, which used to be the most beautiful and oldest in the world, and it is for them that, in the once-excellent restaurants, they program the repugnant and falsified cuisine that the executives always appreciate in loud voices so that the people at the other tables can hear that they have learned their good pronunciations from the announcements on the multi-lingual loudspeakers at airports. "Oh, Plebe! Created worse than all the rest."[147]

Politically, this new class perpetually oscillates, because it successively wants to attain contradictory things. Thus there is not a single political party that does not compete with the others for the executive's vote and, at different times, each one gets it from him.

Like the members of the old petit-bourgeoisie, the executives of today are very diverse, but the strata of upper-level executives, who are the model and illusory goal for all the others, is already tied in a thousand ways to the bourgeoisie properly speaking and it integrates new members into itself much more often than it provides them for itself. There, in a few words, is the

---

[146] English in original.
[147] Dante, *Inferno*, XXXII, 13. Sometimes translated as "O you who are the lowest dregs of all."

portrait of those in whom our bourgeoisie has entrusted a growing portion of its own functions. Thus there cannot be too much reason to be surprised if these functions have been assumed in the bad manner in which they have.

In fact, a progressively growing part of our own class has become parasitical, either through discouragement or inaptitude, and, when this part is not ruined financially, it is at least significantly impoverished, as we might have expected. Well! We will not only say that this part of the bourgeoisie must no longer be defended; we will also say that it must be *eliminated.* Either it will be reintegrated, with dignity and all the intelligence that the current situation requires, into a society whose very tissue we must remake, or, in the opposite case, the Communist ministers who strike that part of the bourgeoisie with a Draconian fiscal reform (one finally worthy of the name "reform") will have our full support. And those comfortable, inactive bourgeois should not believe for a moment that a Communist minister would be necessary to make such a reform, because this measure derives less from the "historic compromise" than their own behavior, which is lacking all combativeness. The people say that necessity sharpens intelligence, and the moment has come in which creativity and the fantastic spirit of enterprise, proof of which the bourgeoisie gave in previous times, today encounters all the conditions for being deployed anew. There are only two possibilities: either the bourgeoisie in Italy and elsewhere proves its intelligence and its will to live, or it will perish, having collaborated too much with its own enemies and thus accelerated and rendered unavoidable its end – because it had wanted to identify its survival as the hegemonic class with the survival of its failings. In that case, its condemnation has already been written:

> For such shortcomings, and not for any other fault, we are lost,
> and are condemned to live here with desire, but no hope.[148]

At the beginning of this final chapter, we alluded to the possibility of making reforms. This is not the place to treat in a profound manner such questions, which we have already envisioned elsewhere, in an unsigned document, very confidentially distributed, entitled *The Republic of the Italians* in homage to a celebrated text by the pseudo-Xenophon.[149] We do not believe

---

[148] Dante, *Inferno*, IV, 40-43.
[149] Pseudo-Xenophon did in fact write a text called *The Constitution of the Athenians,* but it was hostile to its announced subject. As for Censor's *The Republic of the Italians,* it appears that it never existed.

Truthful Report on the Last Chances to Save Capitalism in Italy

we lack modesty when we recall that this document encountered the heartwarming approval of the people who occupy the highest positions of power, because it honors these people that they promptly understood its necessity. Thus we will limit ourselves here to sketching out a few methodological bases for these reforms.

Obviously the difficulty here resides in the necessity of defining what in fact is vital for our economic and social order, that is to say, the necessity of making a severe distinction between those vital things and the appearances that are all too easily accepted by people affected by illusions, readiness and routines. Like everyone else, we recognize that current practices cannot continue, but we do so in a lucid and combative perspective, and not in the imbecilic despondency that currently reigns among all the authors of the errors of the past, who are not even able to discover that they are simply a question of crude errors, with the result that they have the impression that they have been refuted by a thunderbolt from out of the blue, i.e., in a totally unforeseeable manner. In fact, we must correct the irrationalities of our power and, for those who can view our history with disabused eyes, this is nothing new.

Wild capitalism is to be condemned. From the moment that one can sell anything, it is uncivil to only (and with the highest priority) produce what is immediately the most profitable when doing so is detrimental to every conceivable future. All of the excesses of competition must be eliminated by the very power of production, and without delay, because, quite literally, *there is nowhere to live* with this form of production, which destroys its own basis and its own conditions for the future. At a time when the productive process threatens itself because we have believed too much in the value of its *automatism* (which has been helped but never really corrected by political power), all of the social *justifications* for this form of production have universally ceased to be accepted. We no longer believe – no one any longer believes – that the progress of production is capable of *reducing work*. We no longer believe – few people still do – that this form of production is capable of distributing *genuine goods* in increasing quantities and qualities. Thus, conclusions must be drawn. As soon as possible, the true holders of social authority – in the sectors of property, culture, the State and the unions – must secretly, and then publicly, get together to put together a long-term *plan for the rationalization of society*. Capitalism must proclaim and fully realize the rationality that it has carried since its origin, but has only been accomplished partially and poorly. If we can accomplish such urgent and necessary work here in Italy – precisely because our country can draw the strength of salvation from the excesses of the danger – the "Italian model" of capitalism can be adopted by all of Europe and can subsequently open up a new road to the entire world.

From the perspective of a qualitative society, we must, above all, very consciously and clearly distinguish *two sectors* of consumption. One sector should supply authentic quality, with all of its real consequences; the other (that of current consumption) should be cleaned up as much as possible. For a long time, we have feigned to believe that the abundance of industrial production would, little by little, elevate everyone to the conditions of life enjoyed by the *elite*.[150] This argument has so completely lost its very slight appearance of seriousness that, today, it has become degraded to the point of being nothing more than the ephemeral basis for the reasoning and incitements of advertising. Henceforth we must know that the abundance of fabricated objects demands (with ever-greater urgency) the setting up of a true elite, one that precisely shelters itself from such abundance and keeps for itself the little that is really precious. Without this, there will soon be nowhere on Earth where anything truly precious exists. The mechanically egalitarian tendencies of modern industry, which wants to fabricate everything for everyone, and that disfigures and breaks everything that exists so as to distribute its most recent commodities, has spoiled almost all our space and a large part of our time by crowding them both with mediocre goods. Cars and "second homes" are everywhere. If words remain rich, the things they refer to are not, and the landscape is degraded for everyone. The law that dominates here is, of course, that everything that we distribute to the poor can never be anything other than poverty: cars that cannot circulate because there are too many of them; salaries paid in inflated money; meat from livestock fattened up in several weeks by chemical feed, etc.

What would a true *elite*[151] love? Let each reader ask himself this in all sincerity. We love the company of people of good taste and culture, art, the quality of well-chosen food and wine, the calm of our parks and the beautiful architecture of our ancient residences, our rich libraries, and the handling of great human affairs or merely contemplating them from behind the scenes. Who could be convinced that he could have all that, have it be available to everyone else, or only to the top 10 percent of our quite excessively large population, be able to buy it on the market and have it made by our current industries, which produce nothing but cheap junk? And would anyone even

---

[150] French in original.
[151] French in original.

Truthful Report on the Last Chances to Save Capitalism in Italy

dare to suggest that such things can be appreciated and enjoyed by just anyone, even by some guy we have made a government minister but who still feels the sweat of his poor childhood and his feverish arriviste studies?

Thus we must rethink the entirety of production and consumption, and reeducate ourselves in *class consciousness* by reminding ourselves that our class has the historical merit of discovering the existence of socio-economic classes, and that it was the bourgeoisie – not Marxism – that announced the class struggle and founded upon that struggle its possession of society. Our social *elite*[152] is not closed, as were the "states" of the *Ancien Régime*. People have easily gained access to it, over the course of several generations, when our educational system has been realistic and tailor-made for the job, and when we offered to the most suitable individuals the opportunity to enjoy the real advantages that justify the greatest efforts. Likewise, we must remain in a position to offer to the subordinate classes (the craftsmen, the governmental and political/labor union functionaries, etc.) lesser but still satisfying and authentic advantages. Thus, the inclination to valorously elevate oneself on the social ladder so as to attain a qualitatively rich form of existence will be reinforced, because such a goal must appear in all of its beautiful reality and to the precise extent that we can once again begin to enjoy it peacefully. Today, such a reality is out of reach because we have spread false luxury and false comfort so excessively (and without thinking about the consequences) that the entire population is quite unsatisfied by them both.

Miserliness could make the trivial objection that the delimitation of the consumption of things of quality, which would recreate a *barrier of money* against polluted consumption by the lower classes, would also cause unfortunate obligations among the dominant class to spend more money on its everyday purchases. We would respond that the rich must pay for their luxury; otherwise, in a short period of time, they will not have any luxuries at all. The bourgeoisie, especially in Italy, must understand that it is no longer possible for the rich to get everything on the cheap, just as they must also pay their taxes. On the other hand, we must work to improve the people's consumption by correcting, as much as possible, everything harmful to physical or mental health that is currently inflicted on them, and everyone knows that there are a lot of these harmful agents, ranging from our means of transportation to our food, not to mention our mind-numbing distractions and leisure activities. At present, the people are so *worn out* by the abundance of artificial and disappointing consumption that they would accept (with relief)

---

[152] French in original.

consumption that was measured and reassuring, and that pretty much satisfied their authentic needs. It would be sufficient for us – to the extent that we make these corrections – to reveal the reality, especially from the medical point of view, of what has become of bread, wine and the air: in short, all of the people's simple pleasures. If the people are justly *frightened*, we will be praised for having stopped them for sliding any further down the fatal slope of current reality. We must no longer create pollution, except when industry *really* cannot avoid doing so, and then we should only pollute industrial zones that have been set aside and peopled on the basis of fundamental criteria, and not all over the country, *thoughtlessly and casually*,[153] as is done now.

On its own, the question of education is so serious that it would almost suffice to make everyone understand that we must urgently reconstruct a qualitative society, as much in our own well-understood interests as in those of the entire population. When we see the quantity of graduates from what we ironically call our universities, who are not only bereft of real culture but usefulness as well, who cannot even find jobs as workers because employers routinely refuse to hire such people, and who thus inevitably become malcontents and perhaps even rebels, we consider that they are the products of an incompetence that feels no embarrassment in squandering the State's resources, not without result, but, rather, with the result that we are exposed to dangers, and this clashes not only with the most elemental sense of honesty, but with basic logic, too. The Italians – who invented the university and the bank, who during the Renaissance devised the first and best scientific theory of domination – are now the first ones, and more than any other people, to suffer the crisis of everything in which they have excelled. We can still be the world's leaders, that is, if we can show the world the road that will lead us out of and beyond this crisis.

If we offer each person a relatively satisfying place, but especially if we can assure ourselves, without shilly-shallying, of the collaboration of what we might call *the elites of control*, we will not have difficulty resisting all subversion with a minimum of intelligently selective repression. Because it is certainly not the so-called "Red Brigades" that put our power in danger, and if today the four fanatics who compose them seem to be a danger to the State, and easily escape from its prisons, this is not because the "Red Brigades" are a small but very powerful group, but quite simply because the State has faded to such an extent that anyone can make it seem laughable. When we speak of selective repression, we are talking about defending ourselves against

---

[153] *a bischero sciolto,* an old Florentine expression.

Truthful Report on the Last Chances to Save Capitalism in Italy

something other than them.

Censorship – and here we confess that we must keep our Communist allies on a short leash – is not in keeping with the very spirit of capitalism. Censorship can only be envisioned in our laws and used in practice as a completely exceptional recourse, at least when it comes to books. We must neither overestimate their danger nor allow ourselves to forget about them. For example, in the last ten years, and taking into account all of the democratic countries, it seems to us that an intelligent censorship would only have had to ban three or four books in total. But it would have been necessary to make these books disappear absolutely, by every possible means. We ourselves have not neglected to read them, but we did so while keeping them away from everyone else, as the library at the Vatican does with erotic books. When books of political critique only concern topical details or local incidents, they are out of date even before there has been enough time for them to attract a large number of readers. We have only to pay attention to the very rare books that are able to attract followers over long periods of time and eventually weaken our power. We must assuredly educate ourselves about them. Nevertheless, it should not be a matter of criticizing the authors of such books, but annihilating them. Indeed, we know, but often forget, that the pens of such authors always end up making people take up arms, when the opposite does not take place or until the opposite takes place. We no longer remember who said it the first time, but there exists a significant simultaneity between the inventions of printing and gunpowder. In sum, we must treat the authors of certain books as disturbers of the public peace, as harmful to our civilization, which they do not want to reform, but to destroy. On all the crucial points, we must scrupulously guard against all sentimentality and all pretentions to excessive justifications for our censorship. Otherwise we risk corrupting our own lucidity. We do not manage Paradise, but this world.

As terrible as it is, at the moment that we are writing, the situation in Italy is such that no one can accuse us of having exaggerated the danger and discomfort to the point that we have derived all that assaults us as the universal class from the particular misfortunes of this *servile Italy, place of grief, ship without a pilot in a great tempest*.[154] On the contrary, if we are worried about what has happened and what could still happen in Italy, this is precisely because we know that the crisis is global. Given that capitalist unification is so advanced on the planetary scale, it is global capitalism that we risk driving into the abyss. Italy is no longer what it was for a long time: a

---

[154] Dante, *Purgatory*, VI, 75-77.

backwards province, separated from the modern nations. From this situation came both its misfortune and its peace and quiet. Class power is threatened in Russia as it is in America, but Europe – weak in every aspect – is at the center of the tempest. And all the historical misfortunes of Europe have in common the fact that, at the center of them all, one finds the French. Everything permits us to think that, without them, capitalism would have known a superior development from the qualitative point of view. The attack by Charles VIII broke the Italian commercial republics and, three centuries later, Bonaparte did the same thing to Venice. The French Revolution of 1789 gave free rein to the unlimited programs of the riff-raff, while the bourgeois revolution in England in the 17th century appeared to have founded the city politics that permitted the harmonious development of modern capitalism. Finally, even more recently, while the ideology of commodity abundance appeared capable of calming the discontents of the working classes – although it is true that well-informed observers always doubted the stability of such an equilibrium – it was again the French who, in 1968, dealt that ideology its death blow.

What we confront today is a universal problem and, at the same time, a very old one. Last year, Giovanni Agnelli said that the workers no longer want to work because they have been demoralized by the modern living conditions that we have constructed for them. Whatever subtlety we might recognize in this quite original observation, we must say that Agnelli – by privileging too much the examination of circumstances that are the most characteristic of the current period – did not go to the heart of the matter this time. The workers do not want to work every time they glimpse the slightest opportunity for not working, and they glimpse opportunities of this type every time that economic and political domination is weakened by objective difficulties or by difficulties that follow from our blunders. If we get to the heart of the matter, to never work again was the goal of the *Ciompi* as well as the Communards.[155] Every past society in every era has, in its way, confronted this problem and managed to dominate it, while at present we are the ones who are in the process of being dominated by this problem.

Those of our readers who have recognized us know quite well that at no time in our life have we consented to make a pact with fascism, and that we will not make one with any form of totalitarian bureaucratic management, and

---

[155] The *Ciompi* (wool carders) of Florence revolted and set up a short-lived government in 1378. The Communards were partisans of the Paris Commune (1871).

for the very same reasons. The bourgeoisie must want to remain the historical class *par excellence*. Irrefutable on this point, Karl Marx himself demonstrated very well the error that the bourgeoisie commits when it places its political power in the hands of "Bonapartism."[156] Thus, we are turned towards the future, but not any old future.

To speak the language of our "executants," what will be our "model"? While the most cultivated of our adversaries find the rough outline of their model in Pericles' Athens or pre-Medici Florence – models that they must confess are quite insufficient, but nevertheless worthy of their real project, because they display to the most caricatured degree the incessant violence and disorder that are its very essence – we, on the contrary, designate the Republic of Venice as our model of a qualitative society (a model that, in its time, was sufficient and even perfect). Venice had the best ruling class in history: no one resisted it, nor purported to demand an accounting from it. For centuries, there were no demagogic lies, no troubles (or hardly any) and very little blood was spilled. Venice was *terrorism tempered with happiness*, the happiness of each person *in his proper place*. And we do not forget that the Venetian oligarchy, which relied upon the armed workers from Arsenal during certain moments of crisis, had already discovered the truth that an *elite*[157] selected from among the workers always plays the game of society's owners marvelously well.

To finish up, we will say that, rereading these pages, we have not discovered what pertinent objection a rigorous mind could make to them, and we are persuaded that their truth will generally impose itself.[158]

---

[156] Cf. Karl Marx, *The 18th Brumaire of Louis Bonaparte* (1852).
[157] French in original.
[158] Cf. Jonathan Swift's *A Modest Proposal*, the concluding paragraph of which includes this line: "I can think of no one Objection, that will possibly be raised against this Proposal."

Gianfranco Sanguinetti

Proofs of the Nonexistence of Censor By His Creator

I. Phenomenological

In the last ten years, and taking into account all of the democratic countries, it seems to us that an intelligent censorship would only have had to ban three or four books in total. But it would have been necessary to make these books disappear absolutely, by every possible means (...) It should not be a matter of criticizing the authors of such books, but annihilating them (...) We must treat the authors of certain books as disturbers of the public peace, as harmful to our civilization, which they do not want to reform, but to destroy.

(Censor, *Truthful Report*)

Have you read *The Trumpet of the Last Judgment Against Hegel, the Atheist and the Antichrist*?[159] If you still do not know, I can tell you, under the seal of secrecy, that it is by Bauer[160] and Marx. I truly laughed wholeheartedly as I read it.

(G. Jung, letter to Arnold Ruge, December 1841)

Those who up until now regretted not knowing who the author of the *Truthful Report* was, will now regret what they know. Those who were so scandalized by the anonymity of Censor will now have reason to be even more scandalized. Those who praised Censor because they believed it would be good to be seen by a powerful person will no longer be proud of it. And those who until now have prudently preferred to keep quiet and only take a position after they knew the name of the author will have given the measure of everything that their opportunism (like the fearful hesitation that they believe

---

[159] *Author's note*: *Die Posaune des Jügsten Gerichts über Hegel den Atheisten und Antichristen: Ein Ultimatum* (Leipzig, 1841).
[160] Bruno Bauer (1809-1882) was a philosopher, historian and theologian. Nine years older than Karl Marx, he studied with Hegel, who died in 1831.

Truthful Report on the Last Chances to Save Capitalism in Italy

makes a fortress when they are in a predicament) *lets take place.*

In 1841, under the guise of denouncing Hegel as an atheist, Marx and Bauer wrote and published an anonymous pamphlet that was in fact directed against the Hegelian rightwing but that, due to its tone and style, appeared to come from the metaphysical extreme-right of the time. In reality, the pamphlet showed all the menacing revolutionary traits of which the Hegelian dialectic was the bearer in that period, and it was thus the first document that established the death of metaphysics and the "destruction of all the State's laws" that was the consequence.

Today, it is no longer a matter of demonstrating the atheistic and revolutionary character of the Hegelian dialectic, but a matter of knowing if there exists in the dominant class a strategic thought that is capable of conceiving the prospects for capitalism. I have proved that this thought does not exist. I used the following method. If class power today possesses a thought and a project that deals with the preservation of the dominant order, although they are translated into practice with the misfortunes that we see all around us, what would these things be? Everyone has been able to ascertain that, on every occasion they speak, the representatives of power never say anything that is serious, not even about the affairs that concern them the most. And so one wonders, What do they say to each other when they are far away from the public's eyes and ears?

Thus, in August [1975], under the pseudonym of Censor, I wrote and published 520 copies of the subsequently famous pamphlet *Truthful Report on the Last Chances to Save Capitalism in Italy*. This pamphlet was sent to government ministers, members of parliament, industrialists, union leaders and the journalists who are the most respected by public opinion. This *Truthful Report* immediately aroused great interest and a vast discussion that still continues today.

But on one point, at least, everyone was unanimous, because everyone believed that Censor existed, and they ventured to recognize him in this or that person from the economic or political worlds (everyone from Guido Carli to Cesare Merzagora, from Giovanni Malagodi to Raffaele Mattioli himself, who according to some journalists directed "Operation Censor" from beyond the grave).

All of them were deceived: *Censor does not exist.* And although his world still exists, the class that he represents no longer has the strength to produce a bourgeois of such lucidity and cynicism. Giorgio Bocca wrote: "Here's what makes Censor's pamphlet so exceptionally valuable in certain respects: it is one of the rare, extremely rare examples of right-wing culture that doesn't exist among us or doesn't have the courage to manifest itself." Attributing the *Truthful Report* to Merzagora, Enzo Magri wrote that, "it is assuredly the most cynical politico-economic diagnosis that has ever been made in Italy (...) The logic is made of iron, forceful. Censor's rigorous and pitiless analysis leaves no room for any doubts."

Despite the lucid cynicism of Censor, or perhaps precisely because of it, bankers and financiers have greeted my pamphlet with interest. A good number of government ministers, parliamentary representatives and upper-level State functionaries have courteously thanked its first publisher. Some journalists have not managed to hide their admiration, nor even their stupefaction, because the truth is one of the rare things that is capable of causing them to be surprised and spiteful, but also because Censor, in a single blow, destroyed the house of lies that they had patiently but maladroitly constructed over the course of the last few years – on the crucial question of the bombs of 1969, for example. But how could one pretend that the journalists who were incapable of understanding from whence came the *Truthful Report* could, on the other hand, be capable of understanding what has been happening in this country for years? Or from whence came the bombs of 12 December 1969?

All the same, Giorgio Bocca honestly recognized that "this book says more true and terrible things about the hot autumn and the black conspiracies than all of the revolutionary literature," but by saying so he implicitly admits that he does not know the truly revolutionary publications, because, on 19 December 1969,[161] exactly one week later, I published the truth about the bombs of 12 December.

More irritated than all the others, poor Massimo Rira noted in the columns of the *Corriere della Sera* that "this influential person lets it be clear that he knows important particular facts that reinforce the thesis of a 'State massacre,'" and, with consternation, he lets fly a cry of the heart: "How can we not see a sign of the decadence of the [State's] institutions in this inability [to keep quiet] by those who are committed to serve them in silence?" Enzo Magri adds: "The anonymous author supports the thesis of a 'State massacre.' And the logic is made of iron, forceful." The predicament (sometimes noisy, sometimes silent) into which the book has plunged the Italian ruling class and all the political parties is complete and distressing. In the case of "Operation

---

[161] *Author's note*: "Is the Reichstag Burning?" (Milan). [*Translator*: written by Eduardo Rothe and Puni Cesoni, and issued in the name of the Italian section of the Situationist International, of which Sanguinetti was a member.]

Truthful Report on the Last Chances to Save Capitalism in Italy

Censor," there is no doubt that the owners of the social spectacle have, in their turn, been the victims of appearances.

Here are a few other examples of this "phenomenology of error."

"Censor (...) is an enlightened and well-bred conservative, a great tutor of the bourgeoisie, a delegate of private capital (...) Reading this book, we can divine many things concerning Censor's identity." (Carlo Rossella, *Panorama*)

"This pamphlet is certainly a beneficial provocation, an 'Enough!' declared to progressive unction (...) An authentic event, a novelty in which we must rejoice, in the name of culture, even if we aren't in agreement." (*Europa Domani*)

"Who is Censor? (...) His liberal philosophy, his penchants for contempt and reprimanding the politicians, as well as the haughty character of a great bourgeois possessing a very vast experience in the economic domain, emanate from every page of his writing." (Enzo Magri, *L'Europeo*)

"Censor made his *Truthful Report* known in the worst conditions: [only] 520 copies in all, published by a first-time editor, and distributed in the middle of August. And yet its success was immediate. Perhaps because the thesis of the author appeared suggestive to many." (*L'Espresso*)

"Despite his 'conservatism,' Censor casts a benevolent eye upon the Communists and the historic compromise, believing that these new political stabilizers will serve to keep capitalism standing." (*Corriere d'Informazione*)

"Published a few months ago in a numbered edition, this lampoon was immediately reprinted in a commercial edition. But it is both just and unjust, because it is both rare and precious, and thus unusual in publishing; on the other hand, it is exemplary, like a model that merits being proposed to a much larger audience (...) Censor constitutes a political party all by himself: he could be the true gentleman of old minting whose cultural tastes and economic interests are combined in his life, but always safeguarding his decency of life and thought, with a style of comportment and a morality that are true." (Vittorio Gorresio, *La Stampa*)

"Reading [it] reveals a conservative of vast and very refined culture (...) We would like to know more: we would like to have proof of everything that this anonymous person claims. And, until then, we believe that Censor himself has a debt to pay to public opinion: to help it obtain the proof; to speak clear to the bottom without limiting himself to throwing a paving stone into the pool." (Gianna Mazzaleni, *Il Resto del Carlino*)

II. Ontological

Today, the first duty of the press is to undermine the bases of the established political order.

(Karl Marx, *New Rhineland Gazette*, 14 February 1849)

I think of our life in Cologne with pleasure! We are not compromised. That is the essential thing! Ever since Frederick the Great, no one has treated the roguish German people like the *New Rhineland Gazette*.

(Georg Weerth, letter to Marx, 28 April 1851)

Naturally, Marx and Bauer's anonymous pamphlet created a scandal, but after a few weeks its "rightist" provenance was placed in doubt, and its authors' subversive imposture appeared in all its menacing reality. A century and a half later, six months has not been sufficient for Italy to perceive Censor's nonexistence and thus his personal emancipation from metaphysics.

Just as Saint Anselm[162] claimed to provide ontological proof of the existence of God by considering that, if a Being of infinite perfection was conceivable, then it was not inconceivable that this Being could fail to have the fundamental attribute of existence. In the same way, but a millennium later, the Italian bourgeoisie candidly believed that a bourgeois as perfect as Censor

---

[162] Anselm of Canterbury (1033-1109) was the author of the *Proslogion* ("The Discourse on the Existence of God").

Truthful Report on the Last Chances to Save Capitalism in Italy

– since he had all the qualities that it lacked (sincerity, rationality, culture, etc.) – could not fail to have the attribute of existence and, due to that attribute, could contribute to the bourgeoisie's salvation.

Why did our decadent bourgeois so easily believe in the existence of an ally such as Censor? It is quite simple. They believed in it *because they needed to*. And yet, in the words of Vittorio Gorresio, "the only person who could possibly identify the author of the *Truthful Report* was Raffaele Mattioli, who has unfortunately disappeared." But if conceiving of a bourgeois like Censor obligated the bourgeois to invent him, this is the best proof of the fact that, in our ruling class, there exists no one who can flatter himself with having the qualities that it would like to attribute to Censor.

If we can now, retrospectively, be astonished that, for so many months, none of the people who wrote about Censor publicly expressed any doubts about his existence, it is less surprising to see that many "progressive" bourgeois and a part of the non-Stalinist Left applauded the *Truthful Report* "despite [its author] being a rightist or precisely because he is a rightist," as Giorgio Bocca said. In any case, Censor belonged to a right wing that did not appear more cynical than it really was, but that assuredly *spoke* more cynically than it had ever dared to before. It is in fact sufficient to consider the appalling extremism that the Italian bourgeoisie in its current disarray has accepted and even admired, if one wants to understand the full magnitude of that disarray. Thus, it is worth quoting here several passages from the *Truthful Report* that provide its exact measure.

> "Thus we do not seek to prove that contemporary society is *desirable* (…) We say that *this society suits us because it exists* and we want to maintain it to maintain our power over it." (Preface)

> "Today, from the point of view of the defense of our society, there only exists a single danger in the world, and it is that the workers succeed in *speaking to each other* about their conditions and aspirations *without any intermediaries.* All the other dangers are attached to, or even proceed directly from, the precarious situation that places before us this primary problem, which in many respects is concealed and unacknowledged." (Preface)

> "(…) we will lose all of our reasons for managing a world in which our objective advantages have been suppressed (…) Capitalists

must not forget that they are also human beings, and as such they cannot accept the uncontrolled degradation of *all* human beings and thus the personal conditions of life that they especially enjoy." (Preface)

"All of the historically dominant forms of society have been imposed on the masses, who quite simply must be *made to work*, either by force or by illusion. The greatest success of our modern civilization is that it has been able to place an incomparable *power of illusion* at the service of its leaders." (Chapter I)

"This society produces more and more things to watch. Some people have asked us, moved by perfectly irrelevant sentimentality: 'Must we also love this society?'" (Chapter I)

"Our workers have in no way decided upon what they produce. And this is quite fortunate, because we might wonder what they would decide to produce, given what they are. It is quite sure, whatever the infinite variety of conceivable responses, that a single truth would be constant: they would assuredly not produce anything suitable for the society that we manage." (Chapter I)

"Because one must be able to choose between two equivalent commodities, one must also be able to choose between two representatives." (Chapter I)

"Of those minds and hearts that have become discouraged because, for the last ten years, they have taken the end of the troubles of a particular time for the end of the time of troubles, we ask, 'Must we be resigned to the idea that any certainty that has been triumphantly conquered will be ceaselessly put into question, and is the crisis in society destined to always last?' We will respond coldly, 'Yes.' (...) Our world *is not made for the workers,* nor for the other strata of impoverished salaried workers whom our reasoning must place in the simple category 'proletarian.' But every day our world must be made *by* them, under our command. This is the fundamental contradiction with which we must live." (Chapter I)

Truthful Report on the Last Chances to Save Capitalism in Italy

"And precisely because we dare to admit that the Italian workers, who have taken the offensive in the social war, are our enemies, we know that the Communist Party is our support." (Chapter VI)

"Because we should not forget for an instant that the workers, at least when they work and do not revolt, are the most useful reality in the world and merit our respect, for in a certain way they (under our well-informed direction) produce our wealth, i.e., our power." (Chapter VII)

"Henceforth we must know that the abundance of fabricated objects demands (with ever-greater urgency) the setting up of a true elite, one that precisely shelters itself from such abundance and keeps for itself the little that is really precious (...) The law that dominates here is, of course, that everything that we distribute to the poor can never be anything other than poverty: cars that cannot circulate because there are too many of them; salaries paid in inflated money; meat from livestock fattened up in several weeks by chemical feed, etc." (Chapter VII)

"We (...) designate the Republic of Venice as our model of a qualitative society (...) Venice had the best ruling class in history: no one resisted it, nor purported to demand an accounting from it (...) Venice was *terrorism tempered with happiness,* the happiness of each person *in his proper place.*" (Chapter VII)

We could continue to quote many other truths contained in the *Truthful Report.* These are such simple truths, moreover, that anyone would be obligated to admit them, once they have been spoken aloud, but they are such atrocious truths that, until now, no leader has wanted to do so: these are the truths *of this world,* and if they are not pleasing, it is this world that we must transform. And since no one among all those who wrote long articles on Censor protested against any of these atrocities, all these excellent bastards – in accordance with the principle *he who says nothing, consents* – have accepted them.[163] We must remember this.

---

[163] *Author's note*: These bourgeois and these journalists, who preferred to be scandalized by Censor's anonymity rather than the truths contained in his *Truthful Report,* are in fact the same people who, until now, have not shown

Gianfranco Sanguinetti

If the virtuous admirers of Censor had been intelligent, they would have immediately realized that such a pamphlet could only have been written from the point of view of the social revolution (*cui prodest?*),[164] and if they had been unintelligent, but less deficient and less desperate, they would at least have concluded that Censor, as a bourgeois, was quite imprudent and completely unrealistic, since his central project of reconstituting a ruling elite worthy of the name is quite obviously the most impossible utopia. "Operation Censor," and the unlimited stupidity that it revealed,[165] have shown this in the purest experimental light to anyone who by chance had nourished the slightest illusion on the subject. But all these naïve spokesmen for decadence, upon hearing about an elite, already dreamed that they were a part of it.

---

the least qualms about committing or covering up the crimes and monstrous errors of power, of which cynical Censor, had he existed, would have been ashamed. The nonexistence of Censor, so obvious to anyone who read my pamphlet with a grain of salt, but which no one imagined for so long, thus definitively proves the nonexistence of Italy's political personnel, bourgeois intellectuals and bureaucrats. We knew that the majority of our journalists do not know how to write; now we know that they do not know how to read. No contemporary event has shown these people to be so stupid, and since it is not possible that the Italians themselves are equally so, this is the best proof of the stupidity of *the others* who speak in their place, and thus the Italian proletariat must take its affairs directly into its own hands, so as to not leave for an instant more the monopoly of its government and its words to imbeciles of such appalling incompetence.

[164] Latin for "Who benefits?"

[165] *Author's note*: I would like to make clear that I did not lower myself by using subtlety to deceive the "qualified" public to which I sent the *Truthful Report*. Anyone with an average level of culture would have immediately and very easily recognized that, for example, the letter attributed to Louis XVIII is in fact a very well-known literary fake written by Paul-Louis Courier; the letter attributed to a Russian diplomat is a very recognizable passage from a well-known work by Nietzsche; there are long détournements of Tocqueville, and an entire page of the *Report* was taken from *The Real Split in the International* (Paris: Editions Champ Libre, 1972); or a thousand other obvious flippancies. The last phrase of the *Truthful Report*, in itself, is a properly Swiftian enormity. And yet no one noticed any of this and drew the only possible conclusion.

Truthful Report on the Last Chances to Save Capitalism in Italy

III. Historical

In the hospitality of war
We left them their dead as a gift
To remember us by.

(Archilochus, *Fragments*)

There are times in which one can only dispense contempt sparingly, because of the large number of people who need to receive it.

(Chateaubriand, *Memoirs from Beyond the Grave*)

One should not believe that I was motivated by a particular hostility to Italy: I am an internationalist.[166]

---

[166] *Author's note*: If something can console the Italian intellectuals and politicians for having proved their incompetence, it might be the consideration that, in this case, their police are even worse. Some time before giving the manuscript of the *Truthful Report* to the printer, I was released from prison, where I had been thrown, in March 1975, on the extravagant charge of possessing a stockpile of weapons of war, a stockpile whose ghostly existence had never been found except in the completely fantastic enunciation of the accusation against me. This arbitrary act at least allowed the police to conduct four successive searches of both of my residences, and the ones who were in charge found nothing of note in the manuscript, then partially completed, which they read with indiscrete stupidity. At the time, a directive from the Minister of the Interior had orchestrated (in almost every newspaper, including the Stalinist ones and those published by their Leftist imitators) a campaign of calumnies that presented the Situationist International as the hidden power – simultaneously anarchist and fascist – that was organizing terrorism in all of Italy. I am honored to have been a member of the SI, which, by completely different means than terrorism, had unleashed into the world a more authentic and vaster subversion. But it turns out that the SI was dissolved in 1972, due to the very fact of the success of its

Gianfranco Sanguinetti

What did I propose to do by writing such a book and inventing such a person? I proposed to harm Italian capitalism, which is the weakest and most stupid element of class domination in the world, and, more particularly, to harm all those who are engaged in the unfortunate enterprise of rescuing it: the neo-capitalist bourgeoisie and the so-called Communist Party.

Who could be served by such a *Truthful Report*? This is something that no one wondered. As the article devoted to the pamphlet in *Il Borghese* showed, it could only harm the Right. For the Christian Democrats and the other bourgeois governing parties, "Operation Censor" has been even more unfortunate than their enormous errors and brazen provocations because the *Truthful Report* definitively denounced them. For the Stalinist-bureaucratic Left, my pamphlet has been more harmful than a hundred wildcat strikes because it irrefutably demonstrates what the Left's real goals are in Italy today. The enforced silence with which only the press organs of the Italian Communist Party – otherwise so docile in publishing the directives from the Minister of the Interior – have greeted my book is the best proof of this.

In reality, all the political parties have suffered from its publication, because they are all each other's accomplices. But with this operation, the poor Italian State, which has spared us nothing in these last few years – bombs and assassinations that can no longer be counted, although ever since 1969 the workers and almost the entire population have been continuously provoked, deceived and insulted by these crimes, which the bourgeoisie has applauded and about which the Stalinists have cordially kept silent – this *State of provocateurs* has finally been provoked in its turn.

In the *Truthful Report,* there are not only truths, truths that capitalist thought not only does not have the courage to say, but also does not even have the strength to think. Thus, we must wonder: Who does the truth harm? And, Who benefits from the truth? In human history, the truth has always been Public Enemy Number 1 for all power and the principal ally of those who are

---

historical operation, and this dissolution took place at the very moment that the SI had promised to do it: "We will dissolve into the population" (*Internationale Situationniste* #7, April 1962). Moreover, I personally co-signed the act of dissolution with Guy Debord, the author of the well-known book *The Society of the Spectacle* in April 1972 (cf. *The Real Split in the International*). Thus it was perfectly vain to mount such police machinations an entire historical period too late! If they absolutely want to find the situationist critique at work today, they should seek it in the factories held by revolutionaries in Portugal.

exploited. And the Stalinists know these facts better than anyone, because, more than anyone else, they have made a specialty of combating them, in Russia and elsewhere.

What did I want to prove by publishing this pamphlet? Above all, I wanted to prove that the card of the "historic compromise" is the card of the least-backwards capitalism, the one that has enough intelligence to have understood that the so-called Communist Party and the union bureaucracies are its best allies in the permanent social confrontation in which it is opposed to the workers, and this I did not want to demonstrate to the capitalists, who know it all too well due to their experiences, but to the workers. The fact that the bourgeois have taken quite seriously the proposition advanced by Censor that they should conclude the "historic compromise" without any further ado demonstrates the fact that they think that the compromise must be concluded. "Censor is serious," *L'Europeo* wrote, "so serious that his pamphlet can certainly be considered as a real and authentic manifesto of the Italian political and economic right wing." "One immediately understands," *Il Giorno* wrote, "that Censor is serious, and doesn't get lost in the hypocrisies or the bowing and scraping [*les salamalecs*]."

On the other hand, I wanted to prove that the party of social revolution can understand the party of Stalinist-bureaucratic reaction much better than reaction is capable of understanding itself, and I have also proved that the party of reaction can neither understand nor simply recognize the party of revolution, even when it comes forth to do battle.

What the Italian workers are in the process of learning is quite simply what their Portuguese comrades have just learned, what the French revolutionary workers understood in 1968, and what the Russian and Czechoslovakian proletariats (exploited as they are by the vile bureaucratic capitalism that dominates those countries) have always understood: the so-called Communist bureaucrats and unions *are not at all disposed to accept the abolition of the capitalist exploitation of work* in any country in the world. And in Italy, in particular, they are the best servants of our disastrous capitalism, to which they offer their services to spare it from bankruptcy.

In the decline and fall of Italian capitalism, Censor is nothing other than *the reverse image, as in a mirror, of the Italian bourgeoisie,* and the lucid extremism of this nonexistent bourgeois shows the extent and depth of the revolutionary current that invented him. The difference between the two is that, while this revolutionary current *exists,* Censor does not.

The Ministers of the Interior in all the countries, just like the bureaucrats of the so-called Communist parties, feel the same impotent anger about the reappearance of the modern revolutionary movement. In Italy, where the Italian Communist Party hopes to use class struggle as a way of participating in the management of power, and desperately seeks its opportunity, this anger can only be even greater than elsewhere. Because at this point, if revolutionaries can already harm power, which on its own greatly harms itself[, then power is in real trouble]. Look at Portugal: for a year and a half, we have prevented any governmental power from really constituting itself there. The "historic compromise," that Holy Alliance between the bourgeois and Stalinist bureaucrats, which one today proposes to introduce in Italy, has already reigned in Portugal since 24 April 1974: *it reigns but it does not govern.* Pitiful result, ridiculous failure!

What do I want to see happen? The triumph of my party, naturally. And my party is the party of the autonomous organization of workers' assemblies that assume all the powers of decision-making and execution. It is the party of revolutionary workers' councils, the delegates to which are revocable at any moment by the base; the only party that fights all the bourgeois and bureaucratic ruling classes everywhere; the party that, every time it manifests itself, undertakes to realize the abolition of all classes and the State, salaried work and the commodity, and their entire spectacle. And I will never serve any other.

[Gianfranco Sanguinetti]
December 1975

Truthful Report on the Last Chances to Save Capitalism in Italy

Press Clippings

"What does the mysterious Censor say that is so interesting? (...) 'This society suits us because it exists, and we want to maintain it to maintain our power over it.' What society is Censor's? The capitalist society that extends from San Francisco to Vladivostok, the society in which the holders or supervisors of capital succeed in making the masses work by force or by an 'incomparable power of illusion' (...) The last part of the pamphlet is [the product of] an absolute aristocratic cynicism." (*Il Giorno*, 31 August 1975)

"The life and experiences of Censor are intimately tied to those of the most enlightened capitalism in our country." (*Panorama*, 11 September 1975)

"And getting to this point, we wonder who this Censor could be, so involved [as he is] in the secrets of these matters (...) It is thus that what we read further on about the hot autumn, the strategy of tension, and the bombs and massacre at the Piazza Fontana can only be left out [of this review], given the authority that the anonymous writer has already acquired when he reaches this point because of the seriousness of his statements (...) Until now the thesis of the 'State massacre' has only been supported by ultra-Left groups; the Italian Communist Party itself, officially, is quite lukewarm about agreeing with it. But it is stupefying that it is now publicly endorsed by a committed conservative, whose only care is that of saving capitalism in Italy." (*Il Resto del Carlino*, 11 September 1975)

"A small volume with a limited print run theorizes the motivations why large national capital seeks the agreement with the I[talian] C[ommunist] P[arty] (...) Who wrote it is not of great importance, but, on the contrary, the book has such importance from the sole fact that it reflects the ideas of those Italians who believe that the historic compromise will save the bourgeoisie and themselves." (*Il Borghese*, 15 September 1975)

"A real and authentic manifesto of the Italian political and economic right-wing (...) In any case, what is definite is that it is the most cynical political-economic diagnosis ever made in Italy (...) Censor observes that some people will certainly ask of today's [system of] production, 'Must we also love it?' (...)

Gianfranco Sanguinetti

The problem doesn't even have meaning. Because capitalism obviously does not love that system, but only the surplus-value it draws from it." (*L'Europeo*, 18 September 1975)

"A new anonymous author has appeared on the scene of our political literature: he hides himself under the pseudonym 'Censor,' but he doesn't hide his conservative ideas (...) Looks at the Communists and the historic compromise with a benevolent eye." (*Corriere d'Informazione*, 19 September 1975)

"And this is where Censor's anti-conformism manifests itself. Instead of fearing the agreement with the Communist forces, the well-advised bourgeoisie must ally themselves with the ICP so as to utilize its incomparable 'power of illusion' upon the workers for the support of the traditional domination by the merchant bourgeoisie. The true menace against the current stabilizers don't come from the Communist Party, but from the revolutionary possibility of a general rebellion of the masses against their condemnation to salaried work (...) A mystical vision of power, moreover, seems to be the light that guides Censor's thought (...) The psychoanalytic key can no doubt furnish the most fortunate interpretation of the drive that provoked this 'truthful report.' One could speak of the protagonist's complex." (*Corriere della Sera*, 27 September 1975)

"The most recent successful anonymous writer calls himself Censor (...) Incapable of defending itself, the bourgeoisie must conclude a conclude a pact with the ICP to save the capitalist system. But if it doesn't do so immediately, the revolutionary orgy of the proletarians will sweep away the frightened structures of this society." (*L'Espresso*, 5 October 1975)

"We do not share Censor's elitist conception and the aristocratic cynicism that comes from his long familiarity with Machiavelli, Alfieri, Clausewitz and so many conceptual categories from classical literature. We can at least estimate as odd a discourse that is entirely enunciated from the point of view of those who have the real power and the problem of sharing it as least as possible (...) And yet it is a good thing, in all senses, that Censor has proposed a rightist ideological deciphering, a theory of restoration by reforms and suppressions at the point of a sword." (*Europa-Domani*, 15 October 1975)

"It is in sum a perfect construction of very great literary value due to its style,

Truthful Report on the Last Chances to Save Capitalism in Italy

which, by remaining impeccably sustained, doesn't fail to always be amiable, that is to say, accessible (...) Also does justice to the questions that figure on the advertising band placed on the book by its publisher, where we are challenged to divine who Censor is: 'An enlightened conservative? A cynical reactionary? A disguised supporter of the Left?' These are questions that stimulate the curiosity of the reader, but we can tranquilly set them aside, except for the first one and only in part (...) in the sense that the leading lights that he favors prevail over his possible preference for conservatism. His concepts are dialectical, his recommendations are turned towards dynamism (...) and I even find that his constant and precise cultural references testify to a progressive spirit exactly to the extent that culture is progress, without any adjectives." (*La Stampa*, 31 October 1975)

"In a limited number of copies distributed in August, this cynical and refined *Report* has aroused a whirlpool of interpretations (...) Is he a man from the Right or the Left? What does he really want? (...) If someone consciously sought to create a similar success, and if he succeeded, he would be a genius." (*Epoca*, 15 November 1975)

"Censor (...) is so political that it makes us think of a 'great delegate' from the Communist Party. This has the appearance of being a subtle operation by the ICP." (*Il Giorno*, 26 November 1975)

Index

Adelfi, Nicola, 38, 44
Agnelli, Giovanni, ii, 79, 98
Alfieri, Vittorio, 81, 114
Amendola, Giorgio, vi, 37, 75, 77
America (see "United States")
anarchists, ii, 49, 109n
Andreotti, Giulio, vi, 88
Arab-Israeli War of 1973, 62, 66, 67
Athens, 21, 99
Battipaglia, ii, 36
Bauer, Bruno, 100-101, 104
Berlinguer, Enrico, xi, 78
Bocca, Giorgio, 60, 101, 102, 105
Bonaparte, Louis, 46n, 98-99
bourgeoisie, ii, 17, 24, 25, 31, 32, 33, 60, 61, 89, 90, 91, 92, 95, 98, 103, 104, 105, 110, 113, 114
Carli, Guido, vi, 101
Carlyle, Thomas, 12, 15
Cefis, Eugenio, 85, 89
censorship, 96-97, 100
Champ Libre, i, viii, ix, x, 108n
China, xiv, xv, 7, 17
Christian Democratic Party, 26, 38, 39, 70n, 77, 85, 86, 88, 90, 110
Clausewitz, Carl, 45, 52, 54, 66, 114
Communists, iii, 5-7, 24n, 25, 26, 34, 35, 37, 38, 51, 58, 62, 73-80, 82, 84, 86, 103, 114
Constitution, Italian, 29, 47-48, 84
Courier, Paul-Louis, 10n, 108n
Czechoslovakia, xii, 33, 35, 61, 111
Dante Alighieri, 3, 4n, 7n, 31n, 47n, 86, 91n, 92n, 97n
Debord, Guy, i, iii, iv, v, viii, ix, x, xi, 18n, 59n, 65n, 85n, 109n

D'Estaing, Giscard, 69, 70
Donat-Cattin, Carlo, 39, 88
1848 (See "Revolutions of 1848")
Europe, iv, xv, 7, 29, 33, 34, 62, 66, 67, 68-69, 83, 93, 97-98
Fanfani, Amintore, 69, 79, 88
fascism, 15, 20, 24, 25, 60n, 69, 86, 98
FIAT, 40, 53, 79n
Florence, v, 21, 98n, 99
France, ii, iv, v, ix-xi, 30, 33-34, 35, 69, 86
French Revolution, 12, 17, 34, 45, 98
Germany, xi, 33
Gioia, Giovanni, 70, 86
Gladio, xii, 40n, 57n
God, 16, 63, 104
Hegel, Georg Wilhelm, 46n, 50n, 60, 61n, 100
Hegelianism, 75, 101
Il Reichstag Brucia? ii, iii, 49n, 102n
Italian Communist Party (ICP), 6, 37-38, 75-76, 113, 114, 115
Machiavelli, Niccolo, i, 12n, 23n, 50, 62, 68, 69-70, 70n, 73, 75n, 77n, 81n, 85n, 88, 114
Magri, Enzo, 101, 102, 103
Marx, Karl, in, 16, 17, 46n, 67n, 98, 100n, 104
Marxism, 60, 95
Mattioli, Raffaele, 2, 41, 77, 101, 105
May 1968, i, 29-30, 33-34, 70, 98, 111
Miceli, Vito, 40n, 56, 85

Mignoli, Ariberto, iv, v, vi
Milan, ii, v, 36, 40, 42, 43-44, 49n, 54, 102n
Moro, Aldo, vi, 74, 85
NATO, 40n, 57n, 67, 78
Nenni, Pietro, vi, 39
Nietzsche, Friedrich, 61n, 108n
1968 (see "May 1968")
Nixon, Richard, 46, 67, 68
Piazza Fontana, ii, iii, v, vii, viii, xiii, 44, 46, 47, 48, 49n, 56n, 113
Petrarch, 23, 63n, 88n
Plato, 84, 85
pollution, 19, 30, 63, 65, 85, 96
Portugal, 67n, 69, 109n, 112
proletariat, i, ii, 42, 43n, 60, 65n, 87, 90, 107n, 111
Quesnay, François, 16, 17
Reichstag fire (see "Il Reichstag Brucia?")
Retz, Cardinal de, 30n, 32, 54
Revolution, French (see "French Revolution")
Revolutions of 1848, i, 14, 33, 35n, 43, 68n
Rome, iii, 44, 57
Ronchey, Alberto, 75, 89-90

Rumor, Mariano, 37, 39, 55n, 70
Russia, xi, xii, 7, 16, 17, 20, 34, 61, 62, 97, 110, 111
Shakespeare, William, i, 59, 77
SID ("*Servizio Informazioni Difesa*"), 40n, 56, 66
Situationist International, i, ii, iii, iv, v, x, xi, 43n, 49n, 51n, 65n, 66n, 90n, 102n, 109n
Soviet Union (see "Russia")
Spain, i, 69n
Strasbourg, ii, iv
Strategy of tension, iii, xii, 40, 55, 58, 113
Tacitus, 39, 44
terrorism, ii, viii, xi-xiii, xv, 39, 55, 56, 99, 107, 109n
Thucydides, 21, 64
Tocqueville, Alexis de, 35n, 43, 108
Togliatti, Palmiro, 25, 77
United States of America, ii, iv, xi, xiii, xv, 7, 27, 40n, 66, 67-68, 97
Valpreda, Pietro, ii, 49
Venice, Republic of, 99, 107
Welles, Orson, vi-vii

www.ingramcontent.com/pod-product-compliance
Lightning Source LLC
Chambersburg PA
CBHW031633160426
43196CB00006B/404